W9-AEV-192

Carbohydrate–Protein Interaction

Carbohydrate–Protein Interaction

Irwin J. Goldstein, EDITOR
University of Michigan

A symposium sponsored by the

Division of Carbohydrate Chemistry

at the 174th Meeting of the

American Chemical Society,

Chicago, Illinois, August 31–

September 1, 1977.

ACS SYMPOSIUM SERIES **88**

AMERICAN CHEMICAL SOCIETY

WASHINGTON, D. C.   1979

Library of Congress CIP Data

Carbohydrate-protein interaction.
(ACS symposium series; 88 ISSN 0097-6156)
    "Based on a symposium sponsored by the Division of
Carbohydrate Chemistry at the 174th meeting of the
American Chemical Society, Chicago, Illinois, August
31-September 1, 1977."

    Includes bibliographies and index.

    1. Lectins—Congresses. 2. Carbohydrates—Con-
gresses. 3. Proteins—Congresses. 4. Concanavalin A—
Congresses.
    I. Goldstein, Irwin Joseph. II. American Chemical
Society. Division of Carbohydrate Chemistry. III. Series: American Chemical Society. ACS symposium
series; 88.

QP552.L42C37   574.1'9248   78-25788
ISBN 0-8412-0466-7   ASCMC 8   88 1–222 1979

Copyright © 1979

American Chemical Society

PRINTED IN THE UNITED STATES OF AMERICA

ACS Symposium Series

Robert F. Gould, *Editor*

FOREWORD

The ACS SYMPOSIUM SERIES was founded in 1974 to provide a medium for publishing symposia quickly in book form. The format of the SERIES parallels that of the continuing ADVANCES IN CHEMISTRY SERIES except that in order to save time the papers are not typeset but are reproduced as they are submitted by the authors in camera-ready form. As a further means of saving time, the papers are not edited or reviewed except by the symposium chairman, who becomes editor of the book. Papers published in the ACS SYMPOSIUM SERIES are original contributions not published elsewhere in whole or major part and include reports of research as well as reviews since symposia may embrace both types of presentation.

CONTENTS

PREFACE

Although the physiological importance of sugar units in complex carbohydrates was long ignored, studies on carbohydrates as recognition markers in biological processes have blossomed in recent years. Even biologists who did not acknowledge the existence of carbohydrates in biological membranes have accepted the concept of an active role for sugar units in a variety of cellular phenomena which include cellular adhesion, cellular recognition, and density-dependent inhibition of growth.

Enhanced interest in the function of carbohydrate residues derives from the fact that all plant and animal cells are "sugar coated." Because of their strategic position, cell-surface carbohydrates have been implicated in cell–cell communication and in the interaction of cells with their environment.

The possibility that complex carbohydrates like nucleic acids and proteins might also serve as informational molecules has created intensive interest and research on the structure, metabolism, and function of glycoconjugates. One discovery of fundamental importance is that the ordered sequence of sugar units in human cell-surface glycolipids and glycoproteins specify blood-group type. Evidence is also rapidly accumulating to suggest that specific carbohydrate residues on many plasma proteins designate these molecules for uptake by specific cells and tissues. The mechanism involves recognition by membrane-bound receptor (glyco)proteins.

This volume contains the papers presented in a symposium on carbohydrate–protein interaction. The symposium was devoted to an exploration of protein–glycoconjugate interaction in a wide range of biological phenomena: the interaction of enzymes, antibodies, and lectins with complementary carbohydrate molecules; the recognition of carbohydrate-containing structures by chemoreceptors such as taste and other plasma membrane proteins; and the role of carbohydrates in the organization of connective tissue.

Although the physiological function of plant and animal lectins is unknown, these ubiquitous carbohydrate-binding (glyco)proteins can recognize and bind to complex carbohydrates as they occur in solution and on membranes and cell surfaces. A series of papers (by Hardman; Brewer and Brown; Williams and coworkers; Thomas and colleagues; and Evans and Wang) deal with the fundamental chemistry of lectin binding sites. The application of this versatile group of substances as structural and biological probes is described in a paper by Sharon and colleagues, and in a paper by Poretz. Similarly, papers by Pazur and Dreher and by Zopf and colleagues deal with the preparation and carbohydrate-binding specificity of immune antibodies raised in rabbits against sequences of carbohydrate units as they occur in polysaccharides and in synthetic carbohydrate–protein conjugates, respectively.

The stereochemical specificity of taste receptor proteins capable of combining with carbohydrate molecules and of translating the message into sweet sensation is investigated by Jakinovich using a sophisticated neurophysiological technique.

The glycogen debranching enzyme is the first bifunctional eukaryotic enzyme to be reported that consists of a single polypeptide chain. It catalyzes two distinct activities: an oligosaccharide *trans*-glycosylation followed by hydrolysis of an $\alpha(1 \rightarrow 6)$-linked D-glucosyl unit to liberate free glucose. Physical-chemical and kinetic characterization of this novel bifunctional enzyme is described by Nelson.

The molecular mechanism whereby cells recognize, interact with, and internalize glycoprotein molecules is discussed in two papers. Distler and colleagues describe the much studied but controversial role of glycosidically bound D-mannose-6-phosphate in the cellular assimilation of the enzyme β-D-galactosidase; Ashwell and Morgan study the metabolic fate of plasma glycoproteins in fish.

Finally the organization of connective tissue proteoglycans is described in terms of melocular interactions between hyaluronic acid, link protein, and proteoglycan monomers by Rosenberg and colleagues. Support for the molecular structure of proteoglycans is presented in a series of outstanding electron micrographs.

University of Michigan   IRWIN J. GOLDSTEIN
Received September 8, 1978.

Carbohydrate–Protein Interaction

Studies on the Interaction of Lectins with Saccharides on Lymphocyte Cell Surfaces

NATHAN SHARON, YAIR REISNER, AMIRAM RAVID, and AYA PRUJANSKY

Department of Biophysics, The Weizmann Institute of Science, Rehovoth, Israel

Research carried out mainly during the last decade has attested to the importance of the carbohydrate moieties of glycoproteins and glycolipids in biological recognition between cells or between molecules and cells (1-4). Concomitantly, there has been increased activity in the study of carbohydrate-binding proteins such as glycosidases (5), lectins (6-9), anti-carbohydrate antibodies (10), toxins (such as the cholera and botulinum toxins) and glycoprotein hormones (e.g., thyrotropin and human chorionic gonadotropin) (11,12). Investigations of these carbohydrate-binding proteins and their interaction with cells are providing essential clues to the structure of cell-surface sugars and their possible roles in growth, differentiation and development, and in malignant transformation.

I. Properties of Lectins

Lectins are cell agglutinating proteins of nonimmune origin that are widely distributed in nature, being found in plants, microorganisms and animals. They bind mono- or oligo-saccharides with remarkable specificity, in the same way as enzymes bind substrates and antibodies bind antigens. Binding may involve several forces, mostly hydrophobic and hydrogen bonds, and is competitively inhibited by specific sugars.

Many lectins combine preferentially with a single sugar structure, for example D-galactose or L-fucose. For some lectins the specificity is broader and includes several closely related sugars, e.g., D-mannose, D-glucose and D-arabinose; other lectins interact only with complex carbohydrate structures such as those that occur in glycoproteins, glycolipids,

or on cell surfaces. The sugars with which lectins
combine best are those that are typical constituents
of glycoproteins or glycolipids. Perhaps this is a
reflection of the way in which lectins are detected
(namely, by hemagglutination), as a result of which
lectins specific for sugars other than those present
on animal cell surfaces, might be overlooked.

    To date over 50 lectins have been obtained in
purified form (7,9). They vary considerably in amino
acid composition, sugar content, molecular weight,
subunit structure, number of carbohydrate binding
sites per molecule, and metal requirement. Many lectins contain covalently bound sugar and are therefore
glycoproteins. However, concanavalin A, wheat germ
agglutinin and peanut agglutinin, which are among the
best characterized proteins of this class, are devoid
of sugar. Soybean agglutinin, the most thoroughly
investigated glycoprotein lectin, contains six percent
sugar comprised of D-mannose and N-acetyl-D-glu-
cosamine, and consists of four subunits ($M.W.$ 30,000)
each of which carries one carbohydrate chain, D-Man$_9$
(D-GlcNAc)$_2$, linked to the protein via an N-acetyl-D-
glucosaminyl-asparagine linkage. The structure of the
carbohydrate side chain of soybean agglutinin has been
recently elucidated in our laboratory (13). It contains the branched core α-D-Manp-(1→3)-[α-D-Manp-
(1→6)]-β-D-Manp-(1→4)-β-D-GlcNAcp-(1→4)-β-D-GlcNAc,
previously found in many animal glycoproteins as well
as in those from fungi and yeasts and now shown for
the first time to occur in a plant glycoprotein. The
same basic carbohydrate structure has also been found
in the lima bean lectin (14).

    A wide range of specificities and biological
activities has been observed in the interaction of
lectins with cells. Lectins show selectivity in their
agglutination of erythrocytes of different animal
species, with human erythrocytes; some of them are even
blood type specific. Thus, certain lectins agglutinate only human blood type A erythrocytes, while
others agglutinate only type O(H) erythrocytes. Such
lectins are used as an aid in blood typing, especially
since natural anti-O(H) isohemagglutinin in humans is
very rare. Both species and class (T or B) specificity have also been demonstrated in the interaction
of lectins with lymphocytes; moreover, certain
lectins distinguish between lymphocyte subpopulations
from the same animal or organ (see later). Another
intriguing property of lectins is their ability to
agglutinate malignantly transformed cells much better
than normal cells.

Although the role of lectins in nature is not known, there are increasing indications that they function in recognition phenomena of microorganisms, plants and animals, both intercellular and intracellular (for a recent discussion, see 15). They may be responsible for a variety of intercellular interactions, from the adhesion of bacteria to animal cells, to the attachment of sperm to egg. Since recognition also implies distinction between self and nonself and between friend and enemy, these ideas are in line with earlier suggestions that lectins play an important role in host-parasite relationships, both in animals and plants. Lectins may, thus, serve as part of the defense mechanisms of plants against pathogenic microorganisms, whether fungi or bacteria. Recognition by lectins may also be the basis of the association between legumes and their symbiotic nitrogen-fixing bacteria.

The ready availability of lectins, their ease of preparation in purified form, the fact that they are amenable to chemical manipulation and that many of them are inhibited by simple sugars, has made them a most attractive tool in biologic research.

The following is a review of recent studies in our laboratory on the interaction of soybean agglutinin and of peanut agglutinin with lymphocytes from different sources.

II.   Soybean Agglutinin and Peanut Agglutinin

These two lectins have been purified and extensively characterized in our laboratory (reviewed in 7). For the purpose of this presentation, only some of their properties are relevant, the most important of which is their sugar specificity. This specificity is defined in terms of the mono- or oligo-saccharide which inhibits at the lowest concentration the hemagglutinating or precipitating activity of the lectin. Thus, soybean agglutinin is specific for N-acetyl-D-galactosamine, and to a lesser extent (at least 20 times less) for D-galactose (16-18). Peanut agglutinin, on the other hand, is specific for the disaccharide β-D-Galp-(1→3)-D-GalNAc, which is some 50 times better an inhibitor of the lectin than D-galactose (19,20). The sequence β-D-Galp-(1→3)-D-GalNAc is present in many animal glycoproteins, in certain glycolipids, as well as on cell surfaces, e.g., on human erythrocytes and pig lymphocytes. However, this disaccharide is commonly substituted by one or two sialic acid residues, and as a result most

glycoproteins and cells do not interact with peanut
agglutinin. Thus, human erythrocytes or peripheral
lymphocytes do not bind peanut agglutinin to any considerable extent, nor are they agglutinated by this
lectin unless the cells have been treated with
neuraminidase (19,21,22). This is also true for
mouse spleen lymphocytes. As to glycoproteins, peanut
agglutinin does not interact with fetuin or α_1-acid
glycoprotein (orosomucoid) but interacts readily with
asialofetuin and asialo-α_1-acid glycoprotein. A
notable exception are the antifreeze glycoproteins of
antarctic fish, in which unsubstituted β-D-Galp-(1$\rightarrow$3)-
D-GalNAc residues is attached to the polypeptide
backbone; these glycoproteins react with peanut
agglutinin without neuraminidase treatment (23).
There are also certain classes of cells which react
with peanut agglutinin, as will be shown later on.
        Another property of both soybean agglutinin and
peanut agglutinin, which is relevant to this presentation, is the number of sugar binding sites: these
lectins have two sugar binding sites each, per
"monomer" of 120,000 or 110,000 daltons, respectively;
in other words, they are divalent (24,25). In this
respect they differ, for example, from wheat germ
agglutinin and concanavalin A, both of which are
tetravalent at physiological pH.

III. Mitogenic Stimulation

        One of the most striking properties of lectins is
the triggering of quiescent, nondividing lymphocytes
to grow and proliferate, an effect known as mitogenic
stimulation. The gross morphological changes and
biochemical events occurring in lectin-stimulated
lymphocytes in vitro resemble many of the antigen-in-
duced immune reactions that occur in vivo. Lectins
are therefore an important aid in immunology. They
are also used extensively in attempts to understand
the mechanisms by which signals are transmitted from
the outside of the cell to its interior.
        Stimulation of lymphocytes by mitogenic lectins,
such as phytohemagglutinin (PHA) or concanavalin A,
does not require any pretreatment of the cells. This
is not the case with soybean agglutinin, which, as
first demonstrated by Novogrodsky and Katchalski (26)
stimulates mouse spleen cells or human lymphocytes
only after the cells had been treated by neuraminidase.
They suggested that neuraminidase removes from the cell
surface sialic acid residues attached to D-galactose
and N-acetyl-D-galactosamine. The latter sugars are

thus exposed and become accessible to soybean agglutinin; binding to these sites results in stimulation. As expected, this stimulation is inhibited best by N-acetyl-D-galactosamine: at about 10^{-4} M some 50 percent inhibition is observed (27). D-Galactose is a much poorer inhibitor, 10-20 times more of this sugar being required for the same inhibition.

Further studies of the mitogenic properties of soybean agglutinin revealed that different preparations of the lectin exhibited different mitogenic activities, as measured by the concentration of lectin required to give maximal stimulation. The differences observed were often very large, certain preparations being only active at 1-1.5 mg/ml, while others gave a peak of mitogenic stimulation at 10-20 μg/ml. The reason for these anomalous results became clear when it was found that the lectin undergoes aggregation upon storage (28,29). Upon gel filtration of stored (aggregated) soybean agglutinin on Sephadex G-150, three fractions were obtained: the soybean agglutinin monomer (M.W. 120,000) had two sugar binding sites and was not mitogenic up to 1 mg/ml; the tetravalent dimer (M.W. 240,000) exhibited maximal mitogenic activity at 10 μg/ml; and the polyvalent polymer (M.W. $\geq$ 360,000) had the same mitogenic activity as the dimer (27). Soybean agglutinin polymerized by crosslinking with glutaraldehyde is also mitogenic at low concentrations (10-15 μg/ml). More recently, we have found that whereas peanut agglutinin is not mitogenic for neuraminidase-treated mouse lymphocytes (30), the polymerized lectin obtained by crosslinking with glutaraldehyde is mitogenic (31).

Our findings are in agreement with those of Goldstein, Bessler and their co-workers with the lima bean lectins which have a sugar specificity similar to that of soybean agglutinin (32,33). From lima beans several isolectins with different numbers of binding sites for methyl 2-acetamido-2-deoxy-α-D-galactopyranoside have been isolated: a divalent tetramer (M.W. 124,000), a tetravalent octamer (M.W. 247,000) and several higher polymers. The mitogenic activity of the isolectins was related to their valency: whereas the divalent lectin was a poor mitogen, the tetravalent lectin was a strong one.

These findings can be best explained by assuming that in order to induce mitogenic stimulation, the lectin must not only bind to cell surface sugars, but cause crosslinking of cell surface receptors. Such crosslinking may cause lateral movement of transmembrane glycoproteins resulting perhaps in the opening

of a channel for calcium ions. Enhanced movement of
calcium ions across the membrane is, according to one
view, the initial biochemical or metabolic event in
the mitogenic process subsequent to binding of the
lectin to the cell surface (34;   for a discussion,
see ref. 7).

There is indeed evidence that binding of mitogenic lectins causes conformational changes in membrane
structure. We have recently found (31) that when the
Scatchard plots of the binding data of lectins to
lymphocytes were linear, no stimulation of the cells
was observed. However, when the plots were nonlinear,
there was stimulation. The nonlinear plots were
characteristic of binding with positive cooperativity.
In binding phenomena, positive cooperativity implies
that the binding constant of the ligand-receptor complex increases as the extent of occupancy of receptor
site increases. Positive cooperativity in the interaction of lectins with cell surfaces can be explained
by either an increase in the affinity of the receptors
to the lectin, or by an increase in the number of available binding sites caused by unmasking of cryptic
receptors. Both types of change may be the result of
conformational changes in membrane components,
particularly membrane glycoproteins and glycolipids,
the saccharide moieties of which bind to lectins.
Such changes may also be the result of redistribution
of these components in the membrane, facilitated by the
fluid character of the latter.

In order to induce the required conformational
changes, divalence is insufficient, at least for
lectins specific for D-galactose and D-galactose-like
residues interacting with mouse lymphocytes, and these
lectins must be tetravalent (or of higher valence) to
cause the necessary conformational changes in the
membrane. Only such a lectin will pull together the
receptors (with their polypeptide chains and the
transmembrane proteins to which they are attached),
crosslink them and induce clustering of the receptors,
and thus cause the changes in membrane structure that
are prerequisite for mitogenic stimulation.

To obtain further insight into the structure of
the sugars on the cell surface which bind soybean
agglutinin and peanut agglutinin, we used in addition
to neuraminidase the enzyme β-galactosidase from
Diplococcus pneumoniae which removes D-galactose
residues from glycoproteins and from cell surfaces.
Whereas the neuraminidase-treated mouse spleen cells
were stimulated by soybean agglutinin, sequential
treatment with neuraminidase and β-galactosidase

abolished almost completely the stimulation by the
lectin (30). This suggests the presence on the lymphocyte surface of D-galactosyl residues to which soybean
agglutinin binds with ensuing stimulation and makes it
unlikely that N-acetyl-D-galactosamine residues (to
which soybean agglutinin also binds) are involved in
the mitogenic effect. In control experiments with
concanavalin A, stimulation occurred both after incubation with neuraminidase alone and after sequential
treatment with neuraminidase and β-galactosidase.

The above findings on the effect of glycosidases
on lymphocyte stimulation can be readily rationalized
by assuming that the "mitogenic sites" on the cell
surface are part of the commonly occurring aspara-
ginyl-linked sequence α-NANA-(2→6)-β-D-Galp-(1→4)-
β-D-GlcNAcp-(1→2)-α-D-Manp-(1→3,6)-β-D-Manp-(1→4)-
β-D-GlcNAcp-(1→4)-β-D-GlcNAc. As long as the sialic
acid (NANA) residues are present, binding of soybean
agglutinin or peanut agglutinin cannot occur. After
removal of the sialic acid by neuraminidase, soybean
agglutinin and peanut agglutinin bind to the cell
surface, with resultant stimulation. Removal of the
D-galactose residues with β-galactosidase abolishes the
binding to the mitogenic sties. Binding to such sites
of concanavalin A, however, occurs irrespective of the
presence of sialic acid or D-galactose on the carbohydrate side chain since this lectin binds to internal
D-mannose residues.

Another possibility is that the D-galactose-spe-
cific mitogens interact with the O-glycosidically
linked sequence α-NANA-(2→6)-β-D-Galp-(1→3)-α-D-
GalNAcp-(1→0)-Ser(Thr). If this is the case, then
concanavalin A binds to a different oligosaccharide
chain, attached either to the same polypeptide which
carries the O-glycosidically linked moiety (as has been
found, for example, in glycophorin), or to another
polypeptide chain. At present we do not know which is
the sequence involved, nor can we say whether it is
attached to membrane glycoprotein(s) or glycolipids.

One approach to this problem would be to isolate
the intact lectin-receptor molecules from the cell
membranes and characterize them. Work in this direction has been initiated in our laboratory, and a method
for the isolation of the peanut agglutinin receptor
from membranes of neuraminidase-treated human erythrocytes on a column of peanut agglutinin-polyacryl-
hydrazido-Sepharose has been developed (21). The
amino acid composition, D-glucosamine and D-galac-
tosamine content, and the electrophoretic mobility on
polyacrylamide gel electrophoresis in sodium

dodecylsulfate of the peanut agglutinin receptor, were
similar to those of asialoglycophorin. Experiments
using the same approach for the isolation of lectin
receptors from lymphocytes are in progress.

IV. Cell Surface Sugars on Lymphocyte Subpopulations

During recent years it has been established that
lymphocytes are comprised of many subpopulations, such
as T and B, mature and immature cells, etc. Although
various techniques for lymphocyte separation have been
devised, based mainly on cell size, density, electrical
charge and specific surface antigens, none of these is
satisfactory. Studies carried out recently in our
laboratory have demonstrated that differential binding
of lectins to sugars on different cell subpopulations
can serve as a basis for a simple and effective method
for cell fractionation.
The first application of this approach was a
method for fractionation, by the use of peanut
agglutinin, of mouse thymocytes into medullary (immunologically mature) and cortical (immunologically
immature) cells which differ in many of their surface
properties and biological activities (35). As
mentioned earlier, this lectin does not as a rule
interact with cells unless they have been treated with
neuraminidase. However, examination under the microscope of the binding of fluorescein-labeled peanut
agglutinin to mouse thymocytes revealed that the
majority of the cells were stained, whereas a small
proportion (some 10 percent) were not. Moreover, in
contrast to most cells which are not agglutinated by
peanut agglutinin, the bulk of the thymus cells were
agglutinated by the lectin. Separation of the cell
clumps from the single cells was achieved by layering
the agglutinated cell mixture on fetal calf serum (20%)
whereupon the cell clumps settled at the bottom of the
tube and the single, unagglutinated cells remained at
the top. The cell fractions were then collected
separately, suspended in a solution of D-galactose to
dissociate the clumps and remove the lectin, and washed
several times in phosphate-buffered saline.
Examination of the separated cells, which were
obtained in good yield (up to 80 percent), showed that
they were fully viable (>95 percent in each fraction).
In all the properties tested (surface markers, response
to mitogens and immunological activities), the cells
agglutinated by peanut agglutinin were essentially
identical with the immunologically immature cells,
whereas the nonagglutinated fraction consisted of

cells which were similar to the hydrocortisone-resistant mature thymocytes as well as to spleen T cells (35,36). This method has now been used successfully in several other laboratories.

Further experiments have led to the development of a similar method for the separation of mouse spleen T and B cells, although in this case peanut agglutinin could not be used since neither lymphocyte subpopulation binds the lectin, nor are they agglutinated by it. It was found, however, that the B and T cells differ in their ability to bind soybean agglutinin: the B cells were stained by fluorescein-labeled soybean agglutinin, whereas the T cells were not. The spleen lymphocytes could be readily separated by soybean agglutinin, which agglutinated the B cells but did not agglutinate the T cells (37).

On the basis of the results on the binding of peanut agglutinin and soybean agglutinin to the different lymphocyte subpopulations (both before and after neuraminidase treatment), we postulate that these cells carry D-galactose residues (receptors for peanut agglutinin and possibly also for soybean agglutinin) and N-acetyl-D-galactosamine (receptors for soybean agglutinin), which are partially or fully sialylated; the extent of sialylation increases with cell differentiation and maturation. Moreover, it is very likely that peanut agglutinin, which reacts preferentially with early, primitive cells, may serve as a marker for the recognition and identification of such cells. Indeed, peanut agglutinin was found to interact with embryonal carcinoma cells but not with their differentiated derivatives; separation of undifferentiated cells from the differentiated ones, by selective agglutination of the former with peanut agglutinin, was achieved (38).

Another important assumption from the lectin binding studies was that hemopoietic stem cells in the mouse may carry receptors for both peanut agglutinin and soybean agglutinin. Evidence in support of this assumption was obtained when, with the aid of peanut agglutinin and soybean agglutinin a cell fraction was isolated from mouse bone marrow and spleen, which was enriched with hemopoietic stem cells and depleted of "graft-versus-host" activity. Furthermore, such a cell fraction could be used for successful reconstruction of irradiated allogeneic mice (39).

10   CARBOHYDRATE–PROTEIN INTERACTION

Acknowledgments

Studies from the authors' laboratory were supported in part by grants from the Leukemia Research Foundation (Chicago, Illinois), from the Foundation for the Advancement of Mankind (Jerusalem) and by a contribution from a friend of the Weizmann Institute in Argentina. Nathan Sharon is an Established Investigator of the Chief Scientist's Bureau, Israel Ministry of Health.

Literature Cited

1. Sharon, N., Scient. American (1974) 230 (5), 78.
2. Sharon, N., "Complex Carbohydrates - Their Chemistry, Biosynthesis and Functions", 466 pp., Addison-Wesley, Reading, Mass., 1975.
3. Hughes, R. C., "Membrane Glycoproteins", 367 pp., Butterworths, London and Boston, 1976.
4. Sharon, N., Proceedings of the First Congress of the Federation of Asian and Oceanic Biochemists, Nagoya, Japan, Oct. 1977 (in press).
5. Flowers, H. M. and Sharon, N. , Advan. Enzymol. (in press) 48.
6. Sharon, N. and Lis, H., Science (1972) 177, 949.
7. Lis, H. and Sharon, N., In "The Antigens" (ed. Sela, M.), pp. 429, Academic Press. Vol. IV, 1977.
8. Sharon, N., Scient. American (1977) 236 (6), 108.
9. Goldstein, I. J. and Hayes, C. E., Advan. Carbohyd. Chem. Biochem. (1978) 35, 127.
10. Glaudemans, C. P. T., Advan. Carbohyd. Chem. Biochem. (1975) 31, 313.
11. Fishman, P. H. and Brady, R. O., Science (1976) 194, 906.
12. Kohn, L., In "Receptors and Recognition" (ed. Cuatrecasas, P. and Greaves, M. F.), p. 134, Chapman and Hall, London, Series A, Vol. 5, 1978.
13. Lis, H. and Sharon, N., J. Biol. Chem. (1978) 253, 3468.
14. Misaki, A. and Goldstein, I. J., J. Biol. Chem. (1977) 252, 6995.
15. Sharon, N., Proceedings of the Fourth International Symposium on Glycoconjugates, Academic Press (in press).
16. Lis, H., Sela, B. A., Sachs, L., and Sharon, N., Biochim. Biophys. Acta (1970) 211, 582.
17. Pereira, M. E. A., Kabat, E. A., and Sharon, N., Carb. Res. (1974) 37, 89.
18. Hammarström, S., Murphy, L. A., Goldstein, I. J., and Etzler, M. E., Biochemistry (1977) 16, 2750.

19. Lotan, R., Skutelsky, E., Danon, D., and Sharon, N., J. Biol. Chem. (1975) 250, 8518.
20. Pereira, M. E. A., Kabat, E. A., Lotan, R., and Sharon, N., Carb. Res. (1976) 51, 107.
21. Carter, W. G. and Sharon, N., Arch. Biochem. Biophys. (1977) 180, 570.
22. Skutelsky, E., Lotan, R., Sharon, N., and Danon, D., Biochim. Biophys. Acta (1977) 467, 165.
23. Glockner, W. M., Newman, R. A., and Uhlenbruck, G., Biochem. Biophys. Res. Commun. (1975) 66, 701.
24. Lotan, R., Siegelman, H. W., Lis, H., and Sharon, N., J. Biol. Chem. (1974) 249, 1219.
25. Prujansky, A., Lis, H., and Sharon, N., Unpublished results (1977).
26. Novogrodsky, A. and Katchalski, E., Proc. Natl. Acad. Sci. U.S.A. (1973) 70, 2515.
27. Schechter, B., Lis, H., Lotan, R., Novogrodsky, A., and Sharon, N., Eur. J. Immunol. (1976) 6, 145.
28. Lotan, R., Lis, H., and Sharon, N., Biochem. Biophys. Res. Commun. (1975) 62, 144.
29. Lotan, R. and Sharon, N., In "Protein Cross-linking" (ed. Friedman, M.), Part A, p. 149, Plenum.
30. Novogrodsky, A., Lotan, R., Ravid, A., and Sharon, N., J. Immunol. (1975) 115, 1243.
31. Prujansky, A., Ravid, A., and Sharon, N., Biochim. Biophys. Acta (1978) 508, 137.
32. Ruddon, R. W., Weisenthal, L. M., Lundeen, D. E., Bessler, W., and Goldstein, I. J., Proc. Natl. Acad. Sci. U.S.A. (1974) 71, 1848.
33. Bessler, W., Resch, K., and Ferber, E., Biochem. Biophys. Res. Commun. (1976) 69, 578.
34. Maino, V. C., Green, N. M., and Crumpton, M. J., Nature (1974) 251, 324.
35. Reisner, Y., Linker-Israeli, M., and Sharon, N., Cellular Immunol. (1976) 25, 129.
36. Umiel, T., Linker-Israeli, M., Itzchaki, M., Trainin, N., Reisner, Y., and Sharon, N., Cellular Immunol. (1978) 37, 134.
37. Reisner, Y., Ravid, A., and Sharon, N., Biochem. Biophys. Res. Commun. (1976) 72, 1585.
38. Reisner, Y., Gachelin, G., Dubois, P., Nicolas, J.-F., Sharon, N., and Jacob, F., Developmental Biol. (1977) 61, 20.
39. Reisner, Y., Itzicovitch, L., Meshorer, A., and Sharon, N., Proc. Natl. Acad. Sci. U.S.A. (1978) (June issue, in press).

RECEIVED September 8, 1978.

2

The Carbohydrate Binding Site of Concanavalin A

KARL D. HARDMAN

IBM Thomas J. Watson Research Center, Yorktown Heights, NY 10598

Lectins, the class of carbohydrate-binding proteins present in many organisms, have been studied for more than 80 years. The earliest reference commonly quoted for this activity is Stillmark(1) who described the hemagglutinating activity of plant extracts.* Additionally, he later found that *Ricinus communis* (castor beans) contain a toxic protein, ricin, which likewise agglutinates red blood cells. Such proteins have since been of interest to immunologists and more recently to cell biologists for their carbohydrate-binding activity. They provide a variety of useful functions. In fact, the most frequently studied lectin, concanavalin A (Con A)†, is the subject of a book entitled "Concanavalin A as a Tool"(3). Other discussions of the structure and carbohydrate-binding activity of Con A have recently appeared(4,5).

The proteins from the jack bean (*Canavala ensiformis*) were first studied over 60 years ago by Jones and Johns(6). Several years later, Sumner(7), while studying urease (also from the jack bean), isolated three other proteins, two of which could be crystallized, concanavalin A and B. It should be noted that this report of crystalline Con A by Sumner appeared about seven years before he reported the first crystallization of an enzyme, urease(8).

Con A was identified as the hemagglutinin from aqueous extracts of the jack bean by Sumner and Howell(9). Subsequently, the specificities of various plant agglutinins were shown by Boyd and Reguera(10) when they demonstrated that certain seeds contained agglutinins that would react *only* with selected blood group antigens. Con A binds human blood groups very poorly; however, it strongly binds polysaccharides and glycoproteins containing glycosyl residues of the *D-arabino* configuration(11). The monosaccharide which binds most strongly is αMeMan whereas αMeGlc, the C2 epimer of αMeMan binds 1/4 as strongly and galactose, the C4 epimer of glucose, does not bind(11).

---

* For additional references of early work on lectins see Lis and Sharon (ref. 2).
† Abbreviations: Con A, concanavalin A (which refers to the protein with a full complement of manganese and calcium , i.e., the holoprotein); αMeMan, methyl α-D-mannopyranoside; αMeGlc, methyl α-D-glucopyranoside; βIphGlc, β-(0-iodophenyl) D-glucopyranoside; βIphGal, β-(0-iodophenyl) D-galactopyranoside; apo-Con A, metal-free concanavalin A.

0-8412-0466-7/79/47-088-012$05.00/0
© 1979 American Chemical Society

Due to the vast amount of recent research on lectins, it is quite apparent that proteins which bind to specific cell-surface glycoproteins are involved in modulation of a variety of mitotic and metabolic events *within* the cell. It is hoped that structural studies of these proteins and their receptors will prove to be informative in determining the mechanisms of these and other such events. In this presentation, I would like to discuss the structural features of Con A, such as the β-sheets, subunit structure, the manganese, calcium and carbohydrate binding sites and close, by mentioning some recent advances in crystallography which are relevant to the studies of proteins and how these should effect such future research.

The Three-Dimensional Structure

Subunit Composition. The earliest report of molecular weight determination for Con A was by Sumner and coworkers(12) who obtained an average value of 98,000 g/mole at pH 7.3 by ultracentrifugation in Svedborg's laboratory. More recent studies have given values between 50 and 120 K(13-15). Experiments below pH 6 gave values near 50 K, whereas data obtained nearer 7 and above gave higher values. The explanation of these disparate results was clear upon the examination of the first low-resolution electron density maps of crystalline Con A(16). In crystals of native Con A, pH 6.0, space group I222, the molecules are packed as four identical subunits of 26 K daltons each to form pseudotetrahedral clusters of 104 K daltons. However, in solutions below pH 6, these molecules dissociate into dimers. More details of the dimer and tetrameric structures are discussed later. In the native crystals, one asymmetric unit consists of one protomer (chemically identical subunit) which occupies 50% of the volume, the rest consists of solvent. The same subunit structure is found in Con A crystals of the Con A–αMeMan complex at pH 7.4(4), even though in the latter case the space group is different and the tetramers are packed quite differently with respect to their neighbors. The carbohydrate-specific site has been identified in this complex(4) and is discussed later.

Although the explanation is not known, preparations of Con A contain fragmented as well as intact polypeptide chains(17). As many as three polypeptide chains can be isolated, the intact chain of 237 amino acids (molecular weight 26,500), plus two fragments which result from splitting the intact chain between residues 118 and 119(18). Most preparations contain as much as 30% split chains unless special care is taken during preparation. Crystallographic comparison of crystals of purified intact chains and those containing some fragments show no apparent differences in their electron density maps(19). The amino acid sequence has been reported(20) and notable features are that the 40 amino acids at the N-terminus contain a high content of polar side chains whereas the C-terminal half contains an usually high percentage of nonpolar residues, which results in a very low solubility of many proteolytic peptides. The effects of this distribution of amino acid types is quite apparent on examination of the 3-dimensional structure of the monomer. The crystallographic structure has been determined(19,21) and complete atomic coordinates for both studies can be obtained from the Protein Data Bank at Brookhaven National Laboratory.

Monomer. The polypeptide chain of the monomer is outlined in Figures 1, 2A and 2B by vectors linking the α-carbon positions of all 237 amino acids.

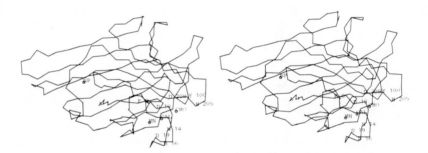

Figure 1.   α-Carbon stereogram of the Con A monomer. Positions marked are for the manganese (MN), calcium (CAL), carbohydrate binding site (CHO), non-polar binding site (NP), and a number of amino acids involved with binding the metal ions and the carbohydrate. The standard single-letter code for amino acids is used. β-sheet I is "behind" the rest of the monomer, while β-sheet II is curved through the center and β-sheet III, which is much smaller, is to the upper-right and connects the first two.

A

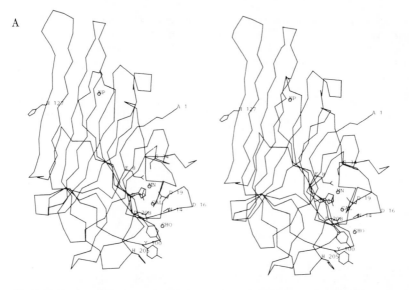

Figure 2A.   Another orientation of the monomer, which includes some of the side
chain atoms in the Mn²⁺, Ca²⁺, and carbohydrate binding sites. β-sheet I is to the
left (including His 127), β-sheet II is through the center (almost perpendicular to
the plane of the paper), and β-sheet III is to the lower-left.

B

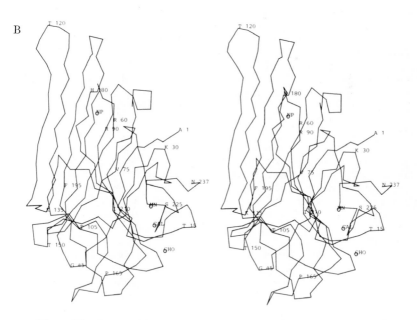

Figure 2B.   Same as 2A, except every 15th residue is labelled for reference

Also shown are the sites for Mn^{2+}, Ca^{2+}, the nonpolar binding site and the carbohydrate-specific site, each occurring once per monomer (discussed later). The most unusual feature of the Con A structure is the arrangement of three β-structure regions in the monomer. The chain folds into a single domain where 50-60% of the amino acids are involved in one of the three β-sheet regions. A high percentage of β-structure was first predicted by Kay(22) from optical rotatory dispersion and circular dichroism studies. Very few other polypeptide chains longer than Con A fold up into a single domain and these (for example, carbonic anhydrase(23) and carboxypeptidase(24)) also have large amounts of β-structure (10 or 8 strands, respectively) twisting through the center of the molecule.

In Con A, the most prominent β-sheet (sheet I) contains 54 amino acids and continues across a crystallographic 2-fold rotation axis, into the second monomer, (Figure 3) forming a continuous 12-strand, flat β-structure "wall"(21). This structural "wall" forms a large outer surface for the dimer and provides the contacts for the tetramer formation (discussed later). The second region, β-sheet II, contains 7 strands which form a semicylindrical "wall" in the center of the subunit. The third region, β-sheet III, is the smallest of the three and is folded across the outer end of the monomer and serves basically as a connection between sheets I and II. This can be seen most readily in Figures 2A and 2B. Sheet I is fairly flat but with a slight twist which is commonly found in other proteins with β-sheets. However, sheet II appears to be quite unique. The upper-inside region of this semicylinder contains polar side chains involved in binding Ca^{2+} and Mn^{2+}, near the carbohydrate side, whereas the lower region contains nonpolar side chains exclusively, which interact with the C-terminal region through numerous van der Waals contacts. The side chains on the "outer" wall of the cylindrical portion are all nonpolar as well and interact with the nonpolar residues of the "inside" of β-sheet I. Details of the hydrogen bonding patterns of these β-structure regions can be found elsewhere(25,26).

Con A contains no standard α-helices, however, there is one loop which contains one or two hydrogen bonds of the α-helical type (residues 80-85), but the carbonyl oxygens appear to tip outward forming bifurcated hydrogen bonds to solvent water molecules. Also, it should be noted that the first 40 amino acids, which have a large percentage of polar side chains, contribute all the atoms which bind directly to the Mn^{2+} and Ca^{2+} ions. This includes the loop which folds around this double ion site (amino acids 10-23) and, in the absence of these ions, would be completely solvated.

Dimer. The two monomers with the most contacts are related across the crystallographic x-axis to form the continuous 12-strand β-sheet (mentioned earlier) and are shown in Figure 3. The intersubunit contacts involve an area of approximately 40 × 25 Å. There are approximately 250 atomic contacts (atoms closer than about 4 Å) of which about 14 are hydrogen bonds(26), the rest are van der Waals contacts. No contacts between this set of monomers would suggest a susceptability to pH changes in the 6 to 7 region. However, in contrast, the intersubunit interface between monomers around the y-axis is larger, about 40 × 40 Å but contains fewer contacts (about 150) and *is* susceptible to pH changes (see below). Furthermore, monomers around the z-axis have almost no contacts.

Across the x-axis (normal to the plane of the paper, through the center of the dimer, Figure 3), 8 amino acids, 124 through 131, form hydrogen bonds to the equivalent strand of the second monomer, by necessity, in an anti-parallel mode (see discussion of the tetramer). A similar feature had been seen earlier in the

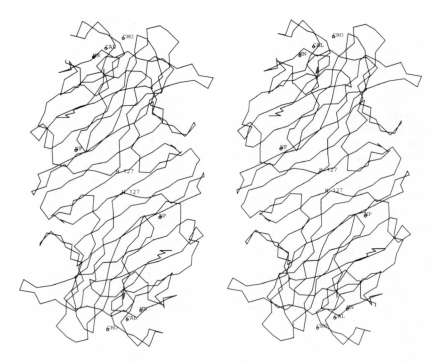

Figure 3.   α-Carbon stereogram of the dimer

structure of insulin(27) where two B chains in the dimer hydrogen-bond across a local 2-fold axis, although in this case there are no additional strands. These similar features were first noted by Dorothy Hodgkin and, since Con A had been observed to mimic the effects of insulin on fat cells(28), the question was raised as to whether these similar structural features were involved with binding to cell surfaces. However, this feature has since appeared more frequently in other unrelated structures, and this correlation seems unlikely.

Tetramer. The center of the tetramer is the intersection of 3 unique 2-fold rotation axes (point group 222). At this point, the 12 strand β-sheets of two dimers interact as mentioned previously (Figure 4). For the dimer these side chains must be on the surface; however, in the tetramer, they must point inward(25). This mode of association does not appear to be simply a function of lattice packing, because in crystals of the Con A–αMeMan complex (space group $C222_1$), three different tetramers are found and *all* have this same orientation(4). The amino acid side chains of β-sheet I which project to the inside of the monomer are entirely hydrophobic, whereas the side chains to the outside of this β-sheet (toward the center of 222 symmetry) are mostly polar. Ten of these per monomer are serine and threonine, while histidine, asparagine, and alanine each contribute two. His 127 is the residue of each monomer nearest the point of 222 symmetry, thus producing a cluster of four imidazole side chains which play a role in the dimer-tetramer association. The apparent mid-point of this transition (dissociation around the y-axis), judged by the data of Pflumm and Beychok(29), is about pH 6.5. Lowering the pH from above pH 7 to below 6 would result in protonation of the imidazole groups and charge repulsion would occur. Also His 51, which interacts with Lys 116 of the second monomer, would affect the dimer-tetramer transition in this pH range. It is also apparent that the center of the tetramer is accessible to solvent molecules even in the crystals, since His 127 and Met 129 bind a number of heavy-atom derivatives without disruption of the crystalline lattice(21,30).

The Carbohydrate Binding Site. It had been shown that βIphGlc binds to a site referred to here as the *nonpolar binding cavity*, Figure 1(31). Since this compound inhibits dextran precipitation by Con A(32), it was assumed to identify the carbohydrate binding site. However, it was then shown that βIphGal and many other nonpolar molecules which do *not* inhibit dextran precipitation also bind to this same site(33). Therefore, the binding of βIphGlc in this cavity is due to the *iodophenyl* moiety and not to the glucosyl residue(33).

Diffusing carbohydrates which specifically bind to Con A, such as αMeMan, αMeGlc or D-glucose, into native crystals (I222) causes disruption of the lattice and the crystals dissolve. Therefore, direct identification of the carbohydrate binding site in *native crystals* is not possible. However, the first suggestion of the correct site was obtained by first covalently crosslinking the crystals with glutaraldehyde and subsequently diffusing in sugars(25). The peaks in the electron density maps were weak since full occupancy could not be obtained and small peaks elsewhere on the surface of the molecule appeared also. Higher concentration of sugars disturbed the lattice of even the crosslinked crystals so these experiments were not considered conclusive.

Subsequently, however, the carbohydrate-specific binding site has been identified in crystals grown from the Con A–αMeMan complex(4). Growth of these crystals (space group $C222_1$) was preceded by incubation of Con A with a large excess of αMeMan. The three-dimensional structure was solved at 6 Å

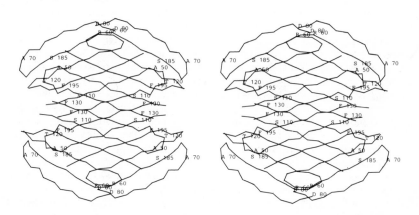

Figure 4.   α-Carbon stereogram of the four β-sheet I sections comprising the tetramer

resolution and the binding site identified by replacing αMeMan by two iodo sugar derivatives, βIphGlc and methyl 2-deoxy-2 iodoacetamido-α-D-glucopyranoside, in different experiments. The atomic coordinates of native Con A were then fit into this low resolution map and the amino acids around the carbohydrate position identified. They are Tyr 12 and 100, Asp 16 and 208, Asn 14, Leu 99, Ser 168 and Arg 228, Figures 5 and 6. This region has a large number of van der Waals contacts and hydrogen bonds to the side chains of Asn 82, Asp 83, Ser 184 and 185 of the neighboring molecule in the I222 lattice. The destruction of the native crystals by the addition of carbohydrate(4) is thus easily explained since there is not sufficient space for a monosaccharide to bind without notably altering these contacts. Additionally, Becker et al.(34) have recently continued studying the carbohydrate binding to cross-linked I222 crystals. They diffused methyl 2-deoxy-2-iodo-α-D-mannopyranoside into the modified crystals and collected data to 3.5 Å. The site they identified as the iodine position was in the same region as mentioned above.

Details of the Mn^{2+}, Ca^{2+}, and carbohydrate binding region are shown in Figures 5 and 6. The orientation of these figures is the same as in Figure 1. The Mn^{2+} and Ca^{2+} are about 4.5 Å apart and form what may be considered to be a double site, since Asp 10 and 19 contribute both carboxyl oxygens, one each as ligands to the Mn^{2+} and Ca^{2+}. They are about 12 and 7 Å from the carbohydrate binding site, respectively. A strand of β-sheet II supplies carboxyls from Glu 8 and Asp 10 and the carbonyl oxygen of Tyr 12 to the Mn^{2+} and Ca^{2+}. The chain then loops out around these ions forming the outer surface of the subunit, from residues 12 to 23, donating Asn 14 and Asp 19 as ligands, then continues back into β-sheet II adding His 24 and Ser 34 (Figure 7). His 24 contributes ligands NE1 to Mn^{2+} and a water molecule appears to bind to both Ser 34 and Mn^{2+}. The sixth coordinate position of the Mn^{2+} appears as a weak region of electron density and must be water of partial occupancy. Therefore, this position is interpreted as the site for the rapidly exchanging water bound to Mn^{2+} found by proton magnetic relaxation dispersion experiments(35). The water between Ser 34 and Mn^{2+} would not be expected to exchange as rapidly and would not have the same symmetry.

The distance from the carbohydrate site to the Mn^{2+} in the crystal is in good agreement with several NMR studies(36-38). These values vary between 10 and 14 Å. From NMR experiments with ^{13}C-enriched α and β anomers of methyl D-glucopyranoside, Brewer et al.(36) calculated average carbon-to-Mn distances for the six non-methyl carbons to be 10.3 and 11 Å, respectively. Similarly, Villafranca and Viola(37) reported an average of 10 Å for the nonmethyl carbons of methyl α-D-glucopyranoside and 13.8 Å for the aglycone methyl carbon from natural abundance ^{13}C-NMR results. Alter and Magnuson(38) reported average ^{19}F - Mn distances of 12.1 and 14.0 Å for the α and β anomers of N-trifluoroacetyl-D-glucosamine at two pH values, 5.1 and 7.0, where Con A is predominantly dimeric and tetrameric, respectively. These results additionally suggest no major differences in carbohydrate binding occur between the dimeric and tetrameric forms(38).

Numerous aryl-pyranosides have been shown to bind more strongly to Con A than their alkyl analogs(32,39,40). Also, acetylation studies with N-acetylimidazole in the presence and absence of carbohydrate have implicated the involvement of tyrosine residues(41). It is quite clear from the three-dimensional structure that Tyr 12 and 100, which are found in this region and exposed to

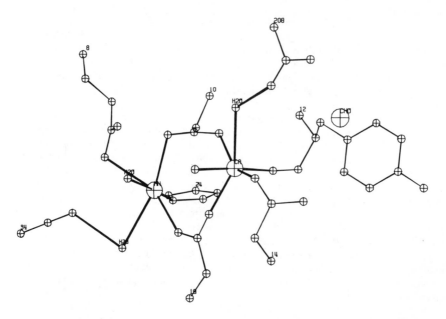

*Figure 5.   The metal ions Mn²⁺ and Ca²⁺ and their ligands, near the carbohydrate
binding site (CHO). (Figures 5–8 have same orientation as Figure 1.)*

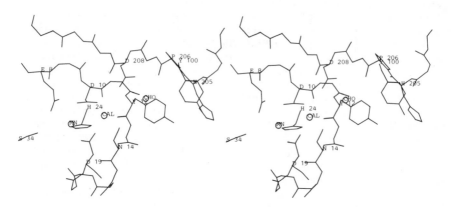

*Figure 6.   Stereogram of the metal ions and carbohydrate binding regions. The
backbone atoms from Glu 8 to Asp 19 and from Ser 203 to Ala 11 plus designated
side chains. The cis peptide bond is between Ala 207 and Asp 208, producing a
distinct "kink" in this strand of β-sheet.*

solvent, could easily account for these increased affinities by a network of π-π interactions.

Chemical modification experiments and hydrogen ion titration studies(42) suggested that two carboxyl groups per monomer are involved in carbohydrate binding. Glu 8, Asp 10 and 19 would appear to remain involved with binding Mn^{2+} and Ca^{2+} and should not be available for reaction. However, Asp 208, which has a water molecule bound between its carboxyl and the Ca^{2+}, is placed suitably for direct interaction with the carbohydrate. Additionally, Asp 16 is located close enough for direct interaction.

Although the location of the binding site has been identified, details of the interaction between the carbohydrate and Con A at atomic resolution are not yet available. NMR results(36,37) have suggested that the plane of the pyranose ring is nearly normal to a line drawn from the carbohydrate to the Mn^{2+} and that C-6 is closer to Mn^{2+} than C-1. The low resolution map of the Con A-αMeMan complex did not allow the identification or orientation of the ring in the site; however, the maximum difference densities for the two iodinated derivatives did suggest that the C-1 carbon extends outward from the protein surface, further from the Mn^{2+} and Ca^{2+} than the rest of the sugar molecule(4). Additionally, the involvement of the Mn^{2+} and Ca^{2+} with carbohydrate binding and the stability of various conformations of the polypeptide chain is evolving as a quite complex situation. This is being studied quite extensively with various NMR techniques and is the subject of another article in this volume(43) as well as elsewhere(44).

The peptide bond between Ala-207 and Asp 208 has been placed in the cis configuration, Figures 6 and 7, in order to obtain a reasonable fit of the polypeptide chain in this region(25). With this exception and perhaps one or two others, only cis peptides involving proline have been found in proteins to date; however, there appears to be no theoretical reason why they cannot occur elsewhere(45). (Proline occurs at 206, one residue before this peptide bond.) Adjacent to this region is Glu 102, which appears to be the only charged side chain on the "interior" of the monomer. These features appear to destabilize the structure in this region and are prime candidates for involvement in conformational changes which occur upon binding carbohydrate(5). It is clear that higher resolution studies of Con A–carbohydrate complexes must be completed to address properly these and similar structural questions.

Di-, tri-, and tetra-saccharides containing α-D-(1→2)-mannosidic linkages have been found to bind much more strongly to Con A than corresponding monosaccharides. If there exists a second pyranosyl binding site adjacent to the first, examination of the three-dimensional model indicates the most likely region involved would be two sections of the polypeptide chain from 97 to 103 and 202 to 208 (Figure 8). These sections include three carboxylic acid side chains, three hydroxyls, two prolines, one lysine, and one histidine. More details of the sugar complexes must await further analyses.

The Con A carbohydrate binding site is quite clearly different from that found in lysozyme, which is a long cleft in the molecule containing up to six subsites for hexopyranose residues(46). The site in Con A is much smaller, perhaps only involving one sugar residue and is not a "groove" but only a shallow indentation in the surface where the pyranose ring lies parallel or almost parallel to the protein surface. Lysozyme has a number of tryptophan side chains involved in binding carbohydrate and in stabilizing the structure around these sites,

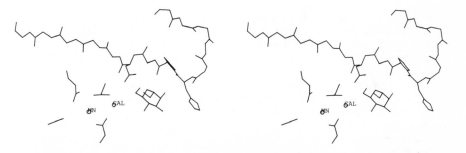

Figure 7.   *The polypeptide backbone from Lys 200 (top-right) to Phe 213. The carbohydrate binding position is represented by the stick drawing of αMeMan. The cis peptide bond is the first peptide bond to the left of Pro 206 (see Figure 6 for the residue numbers).*

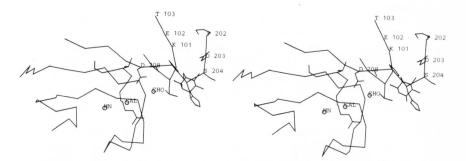

Figure 8.   *Most probable site for the binding of a second pyranoside ring of a polysaccharide (indicated by αMeMan), adjacent to the principal site (CHO) (see text).*

whereas Con A has none. Lysozyme contains very little β-structure and Con A has no α-helix and no disulfide crosslinks. Thus, there are no obvious underlying similarities in the construction of the carbohydrate binding regions, only that the binding involves carboxylic acid side chains and carbonyl oxygens.

The precipitation of large polysaccharides by Con A was first proposed to be analogous to antibody-antigen reactions by Sumner and Howell(9). Since the determination of the three-dimensional structures of Con A and various immunoglobulin molecules and fragments (for a list of references, see ref. 47), structural interests in the lectins has shifted to their specific reactions with carbohydrates with respect to cell surface phenomena, metabolic modulation and cell differentiation. However, there are structural features common to both molecules which are the criss-crossed pairs of β-sheets(21,48), which have also recently been observed in other moleucles(49). In Con A this occurs at two levels: between the 12-strand sheets between dimers and between sheets I and II (Figures 1 and 4). In Con A, prealbumin(50) and the immunoglobulins, binding regions of *two* classes are formed by these pairs of sheets: first, a concave cavity, such as the hapten binding region of immunoglobulin fragments and the center of the Con A tetramer, which bindings a number of heavy-atom molecules; and the second, a convex cavity, such as the nonpolar binding cavity in Con A and the thyroxine binding cavity in prealbumin.

Recent Advances In Crystallography.

In closing, I would like to emphasize three areas of advancement in the field which will soon routinely increase the speed of determining structures of large molecules and greatly improve the resolution and accuracy of the final model. These are: first, the adaptation of two-dimensional detectors for diffraction experiments(51); second, the successful use of cryogenic temperatures for crystallographic enzyme-substrate complexes(52), and third, the development of very fast computational algorithms for refining atomic coordinates, i.e., optimizing large molecular models such as proteins and nucleic acids to crystallographic data(53). It is expected that such advances will significantly advance studies on the correlation of macromolecular structure and biological function.

ABSTRACT

The carbohydrate binding site of concanavalin A has been identified in crystals of the concanavalin A — methyl-α-D-mannopyranoside complex (Hardman and Ainsworth (1976), *Biochemistry* 15; 1120) and is 35 Å from the iodophenyl binding site, which had been postulated to be contiguous with the carbohydrate-specific site (Edelman *et al.* (1972), *Proc. Natl. Acad. Sci. U. S. A.* **69**; 2580 and Becker *et al.* (1975), *J. Biol. Chem.* **250**; 1513). This resolves the disparity between previous interpretations of crystal data and nuclear magnetic resonance data in solution (Brewer, *et al.* (1973), *Biochemistry* 12; 4448). There appear to be no profound conformational changes between the two states. These differences resulted because, in the native crystalline form, o-iodophenyl β-D-glucopyranoside binds as a result of interactions with the iodophenyl moiety, as was shown by the finding that o-iodophenyl β-D-galactopyranoside binds to the identical site (Hardman and Ainsworth (1973), *Biochemistry* 12; 4442). The carbohydrate-specific site is near Tyr 12 and 100, and Asp 16 and 208, a region about 12-14 Å from the manganese ion, in good agreement with NMR studies.

LITERATURE CITED

1. Stillmark, H., Inaugural Dissertation, Dorpat., 1888.
2. Lis, H., and Sharon, N., *Annu. Rev. Biochem.* (1973) 42; 541.
3. Bittiger, H., and Schnebli, H. P., "Concanavalin A As A Tool", Wiley, New York, New York, 1976.
4. Hardman, K. D., and Ainsworth, C. F., *Biochemistry* (1976) 15; 1120.
5. Hardman, K. D., and Goldstein, I. J., in Immunochemistry of Proteins, Vol. 2, Atassi, M. Z., ed., pp. 373-416, Plenum Press, New York, New York, 1977.
6. Jones, D. B., and Johns, C. O., *J. Biol. Chem.* (1916) 28; 67.
7. Sumner, J. B., *J. Biol. Chem.* (1919) 37; 137.
8. Sumner, J. B., *J. Biol. Chem.* (1926) 69; 435.
9. Sumner, J. B., and Howell, S. F., *J. Bacteriol.* (1936) 32; 227.
10. Boyd, W. C., and Reguera, R. M., *J. Immunol.* (1949) 62; 333.
11. Goldstein, I. J., Hollermann, C. E., and Smith, E. E., *Biochemistry* (1965) 4; 876.
12. Sumner, J. B., Gralen, N., and Eriksson-Quensel, I. B., *J. Biol. Chem.* (1938) 125; 45.
13. Olson, M. O. J., and Leiner, I. E., *Biochemistry* (1967) 6; 3801.
14. Agrawal, B. B. L., and Goldstein, I. J., *Biochim. Biophys. Acta* (1967) 147; 262.
15. Kalb, A. J., and Lustig, A., *Biochim. Biophys. Acta* (1968) 168; 366.
16. Hardman, K. D., Wood, M. K., Schiffer, M., Edmundson, A. B., and Ainsworth, C. F., *Proc. Nat. Acad. Sci. U. S. A.* (1971) 68; 1393.
17. Wang, J. L., Cunningham, B. A., and Edelman, G. M., *Proc. Nat. Acad. Sci. U. S. A.* (1971) 68; 1130.
18. Wang, J. L., Cunningham, B. A., Waxdal, M. J., and Edelman, G. M., *J. Biol. Chem.* (1975) 250; 1490.
19. Edelman, G. M., Cunningham, B. A., Reeke, G. N., Jr., Becker, J. W., Waxdal, M. J., and Wang, J. L., *Proc. Nat. Acad. Sci. U. S. A.* (1972) 69; 2580.
20. Cunningham, B. A., Wang, J. L., Waxdal, M. J., and Edelman, G. M., *J. Biol. Chem.* (1975) 250; 1503.
21. Hardman, K. D., and Ainsworth, C. F., *Biochemistry* (1972) 11; 4910.
22. Kay, C., *FEBS Lett.* (1970) 9; 78.
23. Kannan, K. K., Lijas, A., Waara, I., Bergsten, P. C. Lörgren, S., Strandberg, B. Bengtsson, U., Carlbom, U., Fridborg, K., Jarup, L., and Petef, M., *Cold Spring Harbor Symp. Quant. Biol.* (1971) 36; 221.
24. Lipscomb, W. N., Hartsuck, J. A. Reeke, G. N., Jr., Quiocho, F. A., Bethge, P. H., Ludwig, M. L., Steitz, T. A., Muirhead, H., and Coppola, J. C., *Brookhaven Symp. Biol.* (1968) 21; 24.
25. Hardman, K. D., *Adv. Exp. Med. Biol.* (1973) 40; 103.
26. Reeke, G. N., Jr., Becker, J. W., and Edelman, G. M., *J. Biol. Chem.* (1975) 250; 1525.
27. Blundell, T., Dodson, G., Hodgkin, D., and Mercola, D., *Adv. Protein Chem.* (1972) 26; 279.
28. Cuatrecasas, P., and Tell, G. P. E., *Proc. Nat. Acad. Sci. U. S. A.* (1973) 70; 485.
29. Pflumm, M. N., and Beychok, S., *Biochemistry* (1974) 13; 4982.

30.　Becker, J. W., Reeke, G. N., Jr., Wang, J. L., Cunningham, B. A., and Edelman, G. M., *J. Biol. Chem.* (1975) 250; 1513.

31.　Becker, J. W. Reeke, G. N., Jr., and Edelman, G. M., *J. Biol. Chem.* (1971) 246; 6123.

32.　Poretz, R. D., and Goldstein, I. J., *Biochemistry* (1970) 9; 2890.

33.　Hardman, K. D., and Ainsworth, C. F., *Biochemistry* (1973) 12; 4442.

34.　Becker, J. W., Reeke, G. N., Jr., Cunningham, B. A., and Edelman, G. M., *Nature* (London) (1976) 259; 406.

35.　Koenig, S. H., Brown, R. D., and Brewer, C. F., *Proc. Nat. Acad. Sci. U. S. A.* (1973) 70; 475.

36.　Brewer, C. F., Sternlicht, H., Marcus, D. M., and Grollman, A. P., *Biochemistry* (1973) 12; 4448.

37.　Villafranca, J. J., and Viola, R. E., *Arch. Biochem. Biophys.* (1974) 160; 465.

38.　Alter, G. M., and Magnuson, J. A., *Biochemistry* (1974) 13; 4038.

39.　Loontiens, F. G., VanWauwe, J. P., DeGussem, R., and DeBruyne, C. K., *Carbohydr. Res.* (1973) 30; 51.

40.　Bessler, W., Shafer, J. A., and Goldstein, I. J., *J. Biol. Chem.* (1974) 249; 2819.

41.　Doyle, R. J., and Roholt, O. A., *Life Sci.* (1968) 7; 841.

42.　Hassing, G. S. Goldstein, I. J., and Marini, M., *Biochim. Biophys. Acta.* (1971) 243; 90.

43.　Brewer, C. F., Koenig, S. H., and Brown, R. D., (1978) this volume, pp.

44.　Brown, R. D., Brewer C. F., and Koenig, S. H., *Biochemistry* (1977) 16; 3883.

45.　Ramachandran, G. N., and Mitra, A. K., *J. Mol. Biol.* (1976) 107; 85.

46.　Ford, L. O., Johnson, L. N., Machin, P. A., Phillips, D. C., and Tjian, R., *J. Mol. Biol.* (1974) 88; 349.

47.　Davies, D. R., Padlan, E. A., and Segal, D. M., *Annu. Rev. Biochem.* (1975) 44,; 639.

48.　Poljak, R. J., Amzel, L. M., Avey, H. P., Chen, B. L., Phizackerley, R. P., and Saul, F., *Proc. Natl. Acad. Sci. U. S. A.* (1973) 70; 3305.

49.　Richardson, J. S., *Nature* (1977) 268; 495.

50.　Blake, C. C. F., Geison, M. J., Swan, I. D. A., Rerat, C., and Rerat, B., *J. Mol. Biol.* (1974) 88; 1.

51.　Charpak, G., *Nature* (1977) 270; 479.

52.　Petsko, G., and Tsernoglu, D., *Am Crystallogr. Assn.*, National Meeting, Abstracts, (Feb.1977) Asilomar, California.

53.　Agarwal, R. C., *Acta Crystallograph.* (1978) in press.

RECEIVED September 8, 1978.

Binding of Mono- and Oligosaccharides to Concanavalin A as Studied by Solvent Proton Magnetic Relaxation Dispersion

C. FRED BREWER

Department of Pharmacology, Albert Einstein College of Medicine, Bronx, NY 10461

RODNEY D. BROWN, III

IBM Thomas J. Watson Research Center, Yorktown Heights, NY 10598

Interest in the protein concanavalin A (Con A)[1], a lectin isolated from the jack bean (<u>Canavalia ensiformis</u>), derives from its unusual biological properties. In particular, its ability to bind to the surface of both normal and transformed cells has made it a powerful tool for exploring a wide variety of cell-surface related biological effects ([1]). The interaction of Con A with cell-surface membranes is related to the saccharide binding properties of the protein. The saccharide binding specificity of Con A has been shown by Goldstein <u>et al</u>. ([2]) to be directed toward the monosaccharides glucose and mannose, which contain similar hydroxyl group configurations at the C-3, 4 , and 6 positions. The protein binds the α-anomers of these glycosides more strongly than the β-anomers.

Since cell surface carbohydrate determinants occur as oligosaccharides in the form of glycoproteins and glycolipids, it is important to understand the interaction of Con A with these larger complex molecules. Goldstein ([3]) has shown that there exist essentially two classes of oligosaccharides that bind to Con A. The first class of oligomers demonstrates no enhanced binding to the protein relative to monsaccharides; the second class shows enhanced binding. Included in the first class are α-(1→3), α-(1→4), α-(1→6) linked oligosaccharides which contain a non-reducing terminal glucose or mannose residue ([2]); in the second class are α-(1→2)-linked mannose oligomers ([4,5]). The α-(1→2) linked trisaccharide of mannose, for example, has a 20-fold greater affinity constant than methyl α-<u>D</u>-mannopyranoside ([4]). The enhanced binding of such oligosaccharides has prompted speculation ([4]) that the carbohydrate combining site of Con A may bind more than one saccharide residue. Thus, Goldstein and others have suggested that the specificity of Con A binding to oligo- and polysaccharides may involve extended interactions of the protein with several carbohydrate residues.

[1] Abbreviations used: Con A for concanavalin A with unspecified metal content; Ca^{2+}-Mn^{2+}-Con A for concanavalin A containing manganese at the S_1 site and calcium at the S_2 site; Ca^{2+}-Zn^{2+}-Con A for concanavalin A containing zinc at the S_1 site and calcium at the S_2 site; α- and β-MDG for methyl α- and β-D-glucopyranoside; α-MDM for methyl α-D-mannopyranoside; β-IPG for o-iodophenyl β-D-glucopyranoside; β-IPGal for o-iodophenyl β-D-galactopyranoside; α-(1 → 2)-mannobioside for O-α-D-mannopyranosyl-(1 → 2)-D-mannose; α-(1 → 2)-mannotrioside for O-α-D-mannopyranosyl-(1 → 2)-O-α-D-mannopyranosyl-(1 → 2)-D-mannose; NMR for nuclear magnetic resonance; and NMRD for nuclear magnetic relaxation dispersion.

0-8412-0466-7/79/47-088-027$05.00/0

In light of the interest in isolating so-called "Con A receptors" from the surface of a variety of cells, it is of considerable importance to determine the mode of interaction of not only simple monosaccharides but also more complex oligosaccharides with Con A in order to elucidate the complete saccharide binding specificity of the protein. The goal of this paper is to explain the mode of interactions of mono- and oligosaccharides to Con A. We report evidence that the carbohydrate combining site of Con A accommodates only one saccharide residue and that the enhanced binding of certain oligosaccharides can be explained by a statistical argument.

Materials and Methods

Preparation of Con A Derivatives. Ca^{2+}-Zn^{2+}-Con A was obtained from Miles-Yeda. Ca^{2+}-Mn^{2+}-Con A was prepared as previously described (6). Atomic absorption analysis of these two Con A preparations showed essentially equal amounts of the transition metal ion and calcium ions. Sample solutions (0.6 ml) contained Con A at the appropriate concentration in pH 5.60, 0.1N potassium acetate buffer, μ = 1.0 in potassium chloride. The final protein concentration was determined spectrophotometrically using $A^{1\%}_{1cm}$ = 12.4 at 280 nm (7,8).

Saccharides. The α-(1→2) mannose oligosaccharides were gifts from Dr. Irwin Goldstein. The synthesis of o-iodophenyl β-D-glucopyranoside (β-IPG) and o-iodophenyl β-D-galactopyranoside (β-IPGal) will be reported elsewhere. The rest of the saccharides used in this study were obtained from commercial sources.

Relaxation Measurements. Measurements of the magnetic field[2] dependence of the solvent water proton relaxation rate (T_1^{-1}), i.e., nuclear magnetic relaxation dispersion (NMRD), were made by the field cycling method previously described (9,10).

Relaxation Theory. The theory of magnetic relaxation and the procedure used for obtaining the relevant parameters are found in Koenig et al. (6).

Results

The NMRD of solvent water protons in solutions of Ca^{2+}-Mn^{2+}-Con A, Ca^{2+}-Zn^{2+}-Con A and free Mn^{2+} is shown in Fig. 1. T_{1para}^{-1}, the paramagnetic contribution of the bound Mn^{2+} ion of the protein to T_1^{-1}, obtained by correcting the observed NMRD of the Ca^{2+}-Mn^{2+}-Con A solution for the water background (T_{1w}^{-1}) and the diamagnetic contribution of the protein experimentally determined from the NMRD of Ca^{2+}-Zn^{2+}-Con A, is shown in Fig. 2, upper curve. When sufficient α-MDG is added to saturate the saccharide binding sites of Ca^{2+}-Mn^{2+}-Con A, T_{1para}^{-1} is reduced by approximately 25%

---

[2] We indicate magnetic field intensity in units of the Larmor precession of frequency of protons in that magnetic field. The conversion is 4.26 kHz = 1 Oersted = 1 Gauss.

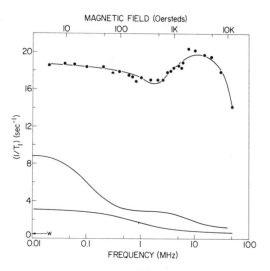

Figure 1. The magnetic field dependence of the spin-lattice relaxation rate of solvent-water protons in solutions of: (top curve) Ca^{2+}-Mn^{2+}-Con A, 1.83 mM (monomer), 1.56 mM bound Mn^{2+}; (middle curve) free Mn^{2+}, 0.15 mM; (bottom curve) Mn^{2+}-Zn^{2+}-Con A, 1.83 mM. The field independent rate for pure water is indicated by W. All measurements were made at 25° in pH 5.60, 0.1 M potassium acetate buffer, $\mu = 1.0$ in KCl. The solid curves are theoretical fits to data.

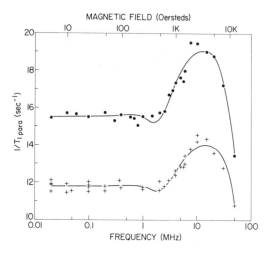

Figure 2. The paramagnetic contribution of the bound Mn^{2+} to the Ca^{2+}-Mn^{2+}-Con A dispersion shown in Figure 1: (top curve) in the absence of sugar; (bottom curve) in the presence of saturating (0.1 M) α-MDG. The solid lines are theoretical fits to the data.

(Fig. 2, lower curve). The lines are from fits to NMRD theory (6). The NMRD of Ca^{2+}-Zn^{2+}-Con A is uneffected by the addition of the saccharide.

When α-MDG is titrated into a solution of Ca^{2+}-Mn^{2+}-Con A, the change in T_{1para}^{-1} at a given magnetic field reflects the fraction F_s of Con A with saccharide bound:

$$F_s = \frac{T_{1para}^{-1}(S_o) - T_{1para}^{-1}(S_T)}{T_{1para}^{-1}(S_o) - T_{1para}^{-1}(S_s)} \quad (1)$$

where $T_{1para}^{-1}(S_o)$, $T_{1para}^{-1}(S_T)$ and $T_{1para}^{-1}(S_s)$ are the T_{1para}^{-1} value in the absence of α-MDG, in the presence of a given total concentration of the saccharide, and in the presence of a sufficient concentration of α-MDG to saturate the carbohydrate binding site of the protein, respectively. Since the water and diamagnetic contributions are essentially independent of sugar concentration at the concentrations used, F_s can be determined from the observed T_1^{-1} values:

$$F_s = \frac{T_1^{-1}(S_o) - T_1^{-1}(S_T)}{T_1^{-1}(S_o) - T_1^{-1}(S_s)} \quad (2)$$

The slope of $1 - F_s$ plotted against $F_s/(S_T - F_s \times P_T)$ (Fig. 3), where P_T is the total Ca^{2+}-Mn^{2+}-Con A concentration, gives the association constant K_a (11). The K_a value determined for α-MDG is 1.1×10^3 M^{-1}, in good agreement with values in the literature (12,13). These results indicate the NMRD spectrum can be used as a monitor of the binding of α-MDG to Ca^{2+}-Mn^{2+}-Con A.

A variety of mono- and oligosaccharides (Table I) were tested for their effects on the NMRD of Ca^{2+}-Mn^{2+}-Con A. Figure 4 shows representative results for several of these saccharides when added in sufficient amounts to saturate the carbohydrate binding sites of the protein. In each case, essentially the same decrease in the NMRD spectrum was observed. In fact, identical results were found for all of the saccharides listed in Table I. β-IPGal which has a low affinity constant did not alter the NMRD at its maximum concentration, 20 mM. Titration of several of the mono- and oligosaccharides such as methyl β-D-glucopyranoside (β-MDG), melezitose and maltotriose in solutions of Ca^{2+}-Mn^{2+}-Con A gave resulting K_a values that agree with previous estimates obtained by other techniques (14). Titration of galactose into a solution of the protein at $25°$ yielded a K_a of ~ 10 M^{-1}, consistent with its weak binding Con A.

Quantitative determination of the effects of binding α-MDG, α-(1→2) mannobioside and α-(1→2) mannotrioside on the NMRD of

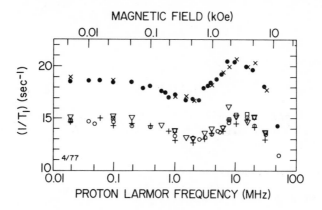

Figure 3.   *Plot for determining the association constant K_a for the interaction of α-MDG with Mn-Con A. F_s, the fraction of Ca^{2+}-Mn^{2+}-Con A molecules with saccharide, is determined from observation of the solvent-proton relaxation rate at 0.04 MHz, as discussed in the text. P_T is the total Ca^{2+}-Mn^{2+}-Con A concentration, mM (monomer), and S_T the total saccharide concentration, mM. The slope of the line through the data give a K_a of 1.1×10^3 M^{-1}. Measurements were made at 25° in pH 5.6, 0.1 M potassium acetate buffer, $\mu = 1.0$ in KCl.*

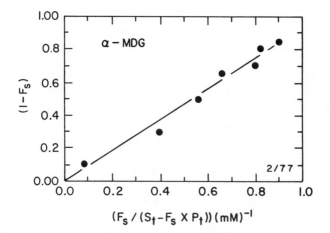

Figure 4.   *The magnetic field dependence of the spin-lattice relaxation rate of solvent protons in solutions of 1.56 mM Ca^{2+}-Mn^{2+}-Con A with no saccharide (●), with 25 mM of β-IPGal (×), and with saturating amounts of: α-MDG, 6 mM (○); α-MDM, 10 mM (□); α-(1 → 2) mannobioside, 83 mM, (▽) and α-(1 → 2) mannotrioside, 15 mM (+). Measurements were made at 25° in pH 5.60, 0.1 M potassium acetate buffer, $\mu = 1.0$ in KCl.*

Table I

List of saccharides used to determine the effects of mono-
and oligosaccharide binding to Ca^{2+}-Mn^{2+}-Con A on the NMRD
profile of the protein.

methyl α-D-glucopyranoside

methyl β-D-glucopyranoside

methyl α-D-mannopyranoside

o-iodophenyl β-D-glycopyranoside

D-galactose

o-iodophenyl β-D-galactopyranoside

maltose

maltotriose

maltotetraose

O-α-D-mannopyranosyl-(1→2)-D-mannose

O-α-D-mannopyranosyl-(1→2)-O-

       α-D-mannopyranosyl-(1→2)-D-mannose

O-α-D-mannopyranosyl-(1→2)-O-α-

       D-mannopyranosyl-(1→2)-O-α-D-mannopyranosyl-

       (1→2)-D-mannose

melezitose

of Ca^{2+}-Mn^{2+}-Con A is shown in Table II in terms of the parameters that determine T^{-1}_{1para}. The NMRD analysis followed the procedure of Koenig et al. ([6]). The parameter which best describes the changes in the NMRD of the protein upon binding of these saccharides is τ_M, the residence time of the exchanging water molecule(s) on the manganese ion, which increases by ~ 40%. A smaller (~15%) change in τ_v is also observed. Changes in the other parameters of the fit (r, τ_{SO} and τ_R) are not considered significant; variation in r and τ_{SO}, are correlated since these parameters enter into the theory as τ_S/r^6. The reason for the anomalously small value for r (~ 2.35 Å) is discussed in Koenig and Brown (to be published).

Discussion

    NMRD measurements of Ca^{2+}-Mn^{2+}-Con A. In an earlier report, Koenig et al. ([6]) observed that the NMRD of Ca^{2+}-Mn^{2+}-Con A was perturbed upon addition of amounts of α-MDG sufficient to saturate the carbohydrate binding sites of the protein. This suggested that changes in NMRD could be used to monitor the binding of saccharides to Con A. Measurements of the T_1^{-1} of solvent water protons have been widely used to obtain information on the binding of organic ligands to paramagnetic metalloproteins ([15]). Koenig and coworkers ([16]) have shown that measurements of T_1^{-1} of water protons over a wide range of magnetic field values (2 Oe to 12 kOe) reveal very different NMRD profiles for the binding of various inhibitors to Mn^{2+}-carboxypeptidase, suggesting different modes of binding in each case. Measurements over this extended magnetic field range permits considerable information to be collected on the effects of ligand interactions with a manganese metalloprotein since the five parameters that contribute to T_1^{-1} of water protons ([6], Koenig and Brown in press) can be evaluated in the absence and presence of bound ligand. Measurements at a single magnetic field may be misleading since the five parameters cannot be evaluated, and different ligands may affect these parameters in different ways ([16]).
    Figure 1 shows the observed NMRD of Ca^{2+}-Mn^{2+}-Con A, Ca^{2+}-Zn^{2+}-Con A and free Mn^{2+} ions in solution over a magnetic field range corresponding to proton Larmor precession frequencies from 10 kHz to 50 MHz. The NMRD for Ca^{2+}-Mn^{2+}-Con A is observed to be distinct from the other two curves. A quantitative analysis of the NMRD of Ca^{2+}-Mn^{2+}-Con A in terms of the parameters that enter into the theory of magnetic relaxation dispersion has been previously published ([6]). Values of these parameters determined from a fit of this theory to the data are given in Table II. Of importance is that all of the manganese ions in the Ca^{2+}-Mn^{2+}-Con A solutions are tightly bound to the protein, and that a water ligand(s) of the manganese ion in the protein is exchanging fairly rapidly (τ_M ~ 6 x 10^{-7}sec)

Table II

Water-Ca^{2+}-Mn^{2+}-Con A interaction parameters[a] from fit of
dispersion theory to data; effect of saturating concentrations
of sugars. Ca^{2+}-Mn^{2+}-Con A = 1.56 mM;   T = 25°;   pH = 5.6.

Sugar	r	τ_V	τ_{SO}
	(Å)	(10^{-1}sec)	(10^{-10}sec)
	(±0.02)	(±0.4)	(±0.05)
NONE	2.30	9.08	1.51
α–MDG	2.29	7.69	1.33
α-(1→2) mannobioside	2.37	7.15	1.75
α-(1→2) mannotrioside	2.43	7.97	1.82

Sugar	τ_M	τ_R
	(10^{-8}sec)	(10^{-6}sec)
	(±0.6)	(±0.02)
NONE	4.88	1.43
α–MDG	6.34	1.96
α-(1→2) mannobioside		1.91
α-(1→2) mannotrioside		1.97

---

[a]

r   Mn-proton distance for the exchanging water.

τ_V Correlation time which describes the magnetic field
               dependence of τ_S.

τ_{SO} Spin-lattice relaxation time of the Mn electronic
               moment   (τ_S at zero magnetic field).

τ_R Rotational correlation time.

τ_M Residence time of the water on the Mn ion.

with bulk solvent to give the observed NMRD profile. There is
a significant contribution of the residence time (τ_M) of the
exchanging water molecule(s) to the observed NMRD of solutions
of Ca^{2+}-Mn^{2+}-Con A (6). The Ca^{2+}-Zn^{2+}-Con A dispersion reflects
a diamagnetic protein with the molecular weight of Con A in
which the bulk solvent "experiences" the Brownian rotational
motions of the protein (10). The dispersion profile of free Mn^{2+}
ions is shown to emphasize the different profiles obtained for
the ion when it becomes bound to Con A along with Ca^{2+} ions.

  Binding of α-MDG to Con A. The magnetic field dependence
of T_{1para}^{-1} in the absence of saccharide is shown in Figure 2,
upper curve. When saturating amounts (0.1 M) of α-MDG are added
to a solution of Ca^{2+}-Mn^{2+}-Con A, the T_{1para}^{-1} dispersion is
reduced at all fields as shown in Figure 2, lower curve. In
order to be sure that this change reflects binding of the
monosaccharide to Con A and not non-specific effects, the
saccharide was titrated into a solution of Ca^{2+}-Mn^{2+}-Con A at 25°
and the incremental changes in T_1^{-1} at 0.04 MHz, were plotted as
discussed above. The results are shown in Figure 3. A K_a
value of 1.1×10^3 M^{-1} was obtained from the plot which agrees
well with values obtained by equilibrium dialysis (12,13). We
thus conclude that the change observed in the NMRD of
Ca^{2+}-Mn^{2+}-Con A upon addition of α-MDG is a result of specific
interactions between the protein and the saccharide.
  A quantitative analysis of the change in the Ca^{2+}-Mn^{2+}-Con A
dispersion upon addition of saturating amounts of α-MDG (Table I)
indicates that the τ_M value of the exchanging water ligand of
the manganese ion of the protein increases upon formation of
the complex (i.e., the residence time of the exchanging water
ligand on the ion becomes longer). Previous studies using
circular dichroism (17) kinetic rate measurements of ^{13}C enriched
α-MDG binding to Ca^{2+}-Mn^{2+}-Con A (18), and recent X-ray crystallographic data (19) suggest a conformational change in the
protein upon saccharide binding. We believe that the change in
the NMRD spectrum of Ca^{2+}-Mn^{2+}-Con A upon binding α-MDG reflects
this conformational change and that the increase in τ_M of the
exchanging water ligand of the manganese ion reflects local
changes in this region of the protein.
  Koenig et al. (6) have previously concluded that the change
in the NMRD of Ca^{2+}-Mn^{2+}-Con A upon addition of α-MDG indicates
that the saccharide does not bind directly to the manganese
binding site in the protein. This conclusion was supported by
the ^{13}C NMR data of Brewer et al. (20,21,22) which show that [13]C
enriched α-MDG binds 10 - 12 Å away from the manganese ion in the
protein-saccharide complex.. Subsequent NMR studies by Villafranca and Viola (23) and Alter and Magnuson (24) confirm these
results. Although earlier X-ray crystallographic studies
reported the carbohydrate binding site to be located 20 Å from
the manganese ion site in the protein (25), recent X-ray

diffraction results (26,27) are now in agreement with the NMR
findings. These latter crystallographic studies show the binding site to be a shallow depression on the surface of the
protein approximately 5 Å from the calcium site and 10 - 13 Å
from the manganese ion site. Furthermore, Becker et al. (19)
have obtained additional results indicating that some groups in
the protein near the saccharide binding site move up to 6 Å
upon saccharide binding to crystalline Con A. The above studies
therefore support our conclusions that changes in the NMRD
profile of Ca^{2+}-Mn^{2+}-Con A upon binding of α-MDG to the protein
reflect a conformational transition in the protein and not
direct binding of the saccharide to the manganese ion.

        Relative Binding Modes of Monosaccharides and Oligosaccha-
rides to Con A. The saccharide binding specificity of Con A was
shown by Goldstein and coworkers (2) to be toward the monosaccharides α-D-mannopyranoside and α-D-glucopyranoside. Their
studies strongly implicated protein-carbohydrate interactions at
the C- 3 , 4 and 6 hydroxyl groups of these saccharides. Using
precipitation-inhibition methods, Goldstein et al. (2) further
established that oligosaccharides containing non-reducing
terminal mannose or glucose residues that were linked through
either α-(1→3), α-(1→4) , or α-(1→6), showed equal binding to Con A
when compared with the corresponding monosaccharides. These
data suggested that Con A interacts with such oligosaccharides
and polysaccharides via their non-reducing terminal carbohydrate
residues. However, Hehre (28) noted that polysaccharides
containing α-(1→2) mannose residues appeared to have internal
residues binding to Con A as judged by their increased agglutination activity with the protein. Further studies by So and
Goldstein (4) showed that α-(1→2) mannose-oligosaccharides
possessed enhanced binding relative to α-MDM: a 5-fold increase
in binding activity for the disaccharide, α-(1→2) mannobioside;
a 20-fold increase in binding activity for the trisaccharide,
α-(1→2) mannotrioside. Other reports indicated that certain
complex glycopeptides showed enhanced binding activity (3,29).
Kornfeld and Ferris (30) reported that a glycopeptide derivative
isolated from an IgE myeloma protein showed a 240-fold increase
in binding relative to α-MDM, the monosaccharide with the
highest known affinity constant for Con A. To account for the
increase binding activity of these oligosaccharides, it was
suggested that the carbohydrate binding site of Con A may bind
to more than one saccharide residue.
        In order to examine the binding specificity of Con A, a
comparative study of the binding of mono- and oligosaccharides
to the protein was carried out using the NMRD profile of
Ca^{2+}-Mn^{2+}-Con A as an index of the conformational change in the
protein induced upon saccharide binding. The rationale for the
experiments follows the observations made by Teichberg and
Shinitsky (31) who observed that oligosaccharides of increasing

length and affinity that bind to lysozyme produce different
conformational changes in the protein, as detected by fluorescence quenching of aromatic residues near the combining site.
As more contacts were made by larger oligosaccharides in the
combining site cleft of lysozyme, which is believed to accommodate up to six saccharide residues, the protein underwent
concomitant steric adjustments. Similar observations have been
made for wheat germ agglutinin which also binds oligosaccharides
(32). By analogy, if oligosaccharides with enhanced binding
activity toward Con A have additional binding contacts with the
carbohydrate combining site, then additional conformational
changes might occur in the protein.

The saccharides tested for their effect on the NMRD profile
of Ca^{2+}-Mn^{2+}-Con A are listed in Table I. They include the
β-anomer of glucose, methyl β-D-glucopyranoside, which
binds a factor of 25-fold less than the α-anomer, and has been
suggested by Brewer et al. (22) to have a different binding
orientation to Con A than α-MDG. α-MDM was also included since
it possesses the highest known affinity constant for Con A for
a monosaccharide. Of the oligosaccharides tested, maltose,
maltotriose and maltotetraose represent a major class of oligomers
which bind to the protein with affinity constants nearly the
same as that of the corresponding monosaccharide, α-MDG. In
addition, several α-(1→2) mannose oligomers were studied since
these oligosaccharides show enhanced binding to the protein.
Melezitose, O-α-D-glucopyranosyl-(1→3)-O-β-D-fructofuranosyl-
(2↔1)-O-α-D-glucopyranoside, has been demonstrated by Goldstein
et al. (2) to show enhanced binding by a factor of approximately
3 relative to α-MDG. The arylglucoside, β-IPG, has been shown
by Brewer et al. (22) to bind to the saccharide binding site of
Con A in solution, but the corresponding arylgalactoside (β-IPGal)
does not. Both of these arylglycosides bind to the "aromatic"
binding site in the protein in the crystalline state (33).
D-Galactose, which binds weakly to Con A, was also tested as
a control.

Representative results of the effect of binding saturating
amounts of several of the saccharides listed in Table I on the
NMRD profile of Ca^{2+}-Mn^{2+}-Con A are shown in Fig. 4. As can be
observed, the results were essentially the same for α-MDG, α-(1→2)
mannobioside, and α-(1→2) mannotrioside. The results, in fact,
were the same for all of the saccharides tested in Table I with
the exception of D-galactose and β-IPGal which bind only very
weakly to the protein.

Table II indicates that τ_M is the parameter which changes
significantly and that this change is the same when α-MDG, α-(1→2)
mannobioside or α-(1→2) mannotrioside binds to Ca^{2+}-Mn^{2+}-Con A.
The data indicate, therefore, that the saccharides listed in
Table I which bind to Ca^{2+}-Mn^{2+}-Con A produce the same change
in the residence time of the exchanging water ligand(s) of the
Mn^{2+} ion of the protein. Since this change in τ_M appears to be

associated with a conformational transition in the protein, the
argument can be advanced that all of the saccharides that bind
to Con A in Table I produce essentially the same conformational
changes in the protein upon binding.

An argument against this conclusion is that our NMRD
measurements of $Ca^{2+}-Mn^{2+}$-Con A are not sensitive to additional
binding interactions that might take place between oligosaccharides such as α-(1→2) mannotrioside, which binds a factor of
20 times better than α-MDM, and the protein. The very small
changes that take place in the circular dichroism spectrum of
the protein upon saccharide binding (17) does not appear to be
a promising way to answer this question. However, another
criterion can be used concerning extended interactions that may
be occurring between the protein and certain oligosaccharides.
In the case of lysozyme and other proteins with extended
combining sites, inhibitors or substrates that bind to these
proteins all possess very similar steric features regarding
their overall conformations. This is demanded by the steric
constraints of the combining sites of these proteins. Therefore,
if Con A possesses an extended combining site, there should be
equal constraints on the three dimensional structures of
saccharides that bind to the protein. However, space-filling
models (CPK) of several of the oligosaccharides that show
enhanced binding such as α-(1→2) mannotrioside, melezitose and
the G1 glycopeptide isolated by Kornfeld and Ferris (30), which
shows enhanced binding, indicate little or no similarity in
their overall conformations. It does not seem likely, therefore,
that these oligosaccharides possess higher K_a values because
of similar extended contacts with the protein. In fact, the
small increments of enhanced binding of factors of 5 and 20 for
α-(1→2) mannobioside and α-(1→2) mannotrioside, respectively,
relative to α-MDM, does not compare with the large enhancements
of $10^2 - 10^3$ in binding that are observed for di- and trisaccharides, respectively, relative to monosaccharides, that bind to
lysozyme (34) which contains an extending binding site. Wheat
germ agglutinin, which appears to have an extended binding site,
also shows large enhancements of $10^2 - 10^3$ for binding di- and
trisaccharides relative to monosaccharides (35). These argu-
ments also weigh against extended interactions between the
binding site of Con A and oligosaccharides.

Our proposed interpretation of the data is that the mono-
and oligosaccharides listed in Table I that bind to Con A all
interact with the protein in the same manner. That is, since
oligosaccharides produce the same change in the NMRD profile as
monosaccharides, their binding modes must be quite similar.
This would suggest that the oligosaccharides are binding
through only one of their residues at any one time, in a manner
similar to monosaccharide interactions with the protein. The
enhanced binding of certain oligosaccharides can therefore be
explained on a statistical basis. Of the oligosaccharides that

possess enhanced binding constants, the α-(1→2) linked mannose
oligomers are most evident. These oligomers differ from the
maltose and isomaltose oligomers, which show no enhanced binding,
in the position of their glycosidic linkages. As previously
noted, Con A demonstrates binding specificity toward saccharides
with glucose or mannose residues that possess free C-3, 4 and
6 hydroxyl groups. Since oligosaccharides with α-(1→3) α-(1→4) or
α-(1→6) glycosidic linkages possess modified hydroxyl groups at
these critical binding positions, only the non-reducing terminal
saccharide of these oligomers can bind to the protein, a
conclusion reached by Goldstein and coworkers (2). However,
where the glycosidic linkage is α-(1→2) for mannose oligomers,
the internal residues also possess free C-3, 4, and 6 hydroxyl
groups. Goldstein et al. (3) has shown that when α-(1→2) mannotrioside was selectively modified at the first and third residues
such that they could not bind to Con A, the resulting derivative
containing an unmodified internal mannopyranoside residue was
observed to bind as well as α-MDM. Using similar derivatization
techniques, the reducing terminal residue of α-(1→2) mannotrioside
was also shown to bind as well as free mannose; Goldstein thus
concluded that internal α-(1→2) linked mannose residues could
bind to Con A as well as non-reducing terminal residues.

Another oligosaccharide with enhanced binding is melezitose,
which is structurally dissimilar to the α-(1→2) mannans. However,
the two are similar in that melezitose also possesses more than
one residue with free C-3, 4, and 6 hydroxyl groups: the first
and third glucose units. The enhanced binding of melezitose by
a factor of seven relative to maltotriose, which contains only
one "binding" residue at the non-reducing terminal end, is
similar to the enhancement observed for α-(1→2) mannobioside
(which also contains two "binding" residues per molecule)
relative to α-MDM. Thus, it appears that a necessary requirement for enhanced binding is for a molecule to contain multiple
glucose or mannose residues which possess free C-3, 4, and
6 hydroxyl groups.

We therefore suggest that the enhanced binding of certain
oligosaccharides to Con A appears to be due not to an extended
binding site on the protein, but rather to effects which result
from clustering several "binding" residues (glucose or mannose)
in the same molecule, any one of which can bind to the
single site on the protein. To our knowledge, the effects of
binding polyvalent ligands to monovalent protein binding sites
has never been studied. Nevertheless, we feel the enhanced
binding could come about from an increase in the forward rate
constant for complex formation which, to the first approximation,
would be expected to be proportional to the number of binding
residues in the molecule. However, the observed enhancements
for α-(1→2) mannobioside and α-(1→2) mannotrioside of 4- and
20-fold, respectively, indicate more than proportional increases
in their binding constants in relation to their number of

"binding" residues. This suggest the possibility of decreased
off-rates for these molecules as well. A possible mechanism for
reducing the dissociation rates of these oligosaccharides may
be one in which such a molecule upon dissociating from the
protein is immediately recaptured by another of its binding
residues. In effect, this would decrease the macroscopic off-
rate of the oligosaccharide by limiting its diffusion away from
the protein once binding has taken place. Such a recapture
mechanism has been observed for the action of α- and β-amylases
on polysaccharide substrates in which so-called multiple attacks
occur on a single substrate oligomer (36,37). We therefore
suggest that a combination of these effects on the forward and
reverse rates of binding of these oligosaccharides could account
for their enhanced binding constants.

        The absolute magnitude of the enhancement effects apparently
reaches a limit between the tri- and tetrasaccharides in the
α-(1→2) mannan series since higher homologs begin to show
decreased apparent binding constants toward Con A (2). However,
this latter decrease may be due to the formation of intermolecular
complexes of the higher oligomers in solution and not an actual
decrease in their binding to Con A.

        This proposed mechanism of enhanced binding of certain
oligosaccharides to Con A requires only a single residue binding
site on the protein. Data which are consistent with this are
the following. Goldstein and coworkers have demonstrated that
a wide variety of monosaccharides including α- and β-MDG, and
oligosaccharides including those with "internal" binding residues
such as O-α-D-galactopyranosyl-(1→2)-O-α-D-mannopyranosyl-(1→2)-
D-mannose competitively displace the chromogenic ligand
p-nitrophenyl β-D-mannopyranoside from Con A. These workers
concluded that the mono- and oligosaccharides used in the study
competed for the same site on the protein. In addition, the
specificity of the same site on Con A would require that all
residues in a saccharide that bind directly to the protein possess
the same molecular specificity as monosaccharides that bind to
the protein. It is well known that the monosaccharides glucose
and mannose bind well, as opposed to galactose which binds
poorly, and that the α-anomers of these monosaccharides bind
better than the β-anomers. Indeed, Goldstein et al. (3) have
shown that when the terminal non-reducing residue of α-(1→2)
mannotrioside is converted to a galactose residue, a loss in
binding is observed. This result is consistent with the terminal
non-reducing mannose residue α-(1→2) mannotrioside binding at the
same site as α-MDM. Furthermore, Goldstein and coworkers (2)
have shown that reduction of mannose to give mannitol eliminates
binding. When the terminal reducing end of α-(1→2) mannotrioside
is correspondingly reduced to give α-(1→2) mannotriitol, a loss
in binding occurs which is also consistent with the terminal
reducing residue of α-(1→2) mannotrioside binding to the same
site as α-MDM. Similar substitutions affect the binding of

another α-(1→2) linked oligosaccharide, kojibiose (o-α-D-gluco-
pyranosyl-(1→2)-D-glucose). Kojibiose shows enhanced binding
by a factor of three relative to o-α-D-galactopyranosyl-(1→2)-
D-glucose (5). This suggests that the non-reducing residue of
kojibiose contributes to binding, and the specificity of binding
is similar to that observed for monosaccharide binding to Con A.
Interestingly, the α-methyl anomer of kojibiose shows a further
enhancement of binding by a factor of two relative to kojibiose,
which is presumed to be a mixture of the α- and β-anomers.
This suggests that the reducing end residue of the disaccharide
binds to the protein at a site which also preferentially binds
the α-anomers of monosaccharides. The above data are all
consistent with the protein containing a single binding site
which can interact with either terminal non-reducing residues,
internal residues, or terminal reducing residues of oligosaccharides providing that these glucose or mannose residues possess
free C-3, 4, and 6 hydroxyl groups.

   Thus, from consideration of all of the available evidence,
it appears that the saccharide binding specificity of Con A can
be accounted for by a binding site which accommodates only one
saccharide residue, and that oligosaccharides containing
multiple glucose or mannose residues which have free C-3, 4, and
6 hydroxyl groups can demonstrate enhanced binding to the
protein, relative to monosaccharides, due to increases in their
probability of binding. These results have important implica-
tions regarding the molecular properties of "so-called" Con A
receptors on the surface cells.

Acknowledgments

   This work was supported in part by grant # CA-16054,
awarded by the National Cancer Institute, Department of Health,
Education and Welfare. C.F. Brewer is a recipient of a Research
Career Development Award, grant # 1-K04-CA-00184 from the Dept.
of Health, Education and Welfare. (A preliminary report of the
work was presented at the University of Oklahoma Symposia on
Concanavalin A, (Brown, R.D., Brewer, C.F. and Koenig, S.H.
(1975), in Concanavalin A, Adv. Exp. Med. Biol., 55, 323).

Abstract

   We have measured the effects of binding of a series of mono-
and oligosaccharides to Ca^{2+}-Mn^{2+}-Con A on the solvent water
proton relaxation rate over a range of magnetic fields from 5 Oe
to 12 kOe. We find that the binding of methyl α- and β-D-
glucopyranoside, methyl α-D-mannopyranoside and o-iodophenyl β-
D-glucopyranoside produce the same increase in the residence
time of the exchanging water ligand(s) of the Mn^{2+} ion and
therefore the same conformational change in the protein, whereas
galactose and   o-iodophenyl β-D-galactopyranoside, which do not

bind under the same conditions, show no effects. The same
reduction in relaxation rate as that caused by the above monosaccharides was observed with the following oligosaccharides:
$\underline{D}$-maltose, $\underline{D}$-maltotriose, $\underline{D}$-maltotetraose, $\underline{O}$-α-$\underline{D}$-mannopyranosyl-
(1→2)-$\underline{D}$-mannose, $\underline{O}$-α-$\underline{D}$-mannopyranosyl-(1→2)-$\underline{O}$-α-$\underline{D}$-mannopyranosyl-
(1→2)-$\underline{D}$-mannose, $\underline{O}$-α-$\underline{D}$-mannopyranosyl-(1→2)-$\underline{O}$-α-$\underline{D}$-mannopyranosyl-
(1→2)-$\underline{O}$-α-$\underline{D}$-mannopyranosyl-(1→2)-$\underline{D}$-mannose and melezitose.
Goldstein and coworkers have shown that the first three oligo-
saccharides have nearly the same affinity as monosaccharides,
whereas the α-(1→2) linked mannans show increasing affinity
constants with increasing chain length. Melezitose also shows
enhanced binding by a factor of three relative to methyl-α-$\underline{D}$-
glucopyranoside. The water relaxation data suggest that the
above mono- and oligosaccharides bind to Con A by a similar
mechanism involving only a single saccharide residue combined
with the protein at one time. The greater affinity of melezitose and the α-(1→2) mannose oligosaccharides appears to be due
to an increase in the probability of binding associated with
the presence of more than one binding residue in the chain and
not to an extended binding site on the protein.

Literature Cited

1. Lis, H. and Sharon, N., Annual Revs. Biochem., (1973),
       42, 541.
2. Goldstein, I.J., Hollerman, C.E. and Smith, E.E.,
       Biochemistry, (1965), 4, 876.
3. Goldstein, I.J., Reichert, C.M., Misaki, A. and Gorin, P.A.J.
       Biochim, Biophys. Acta, (1973), 317, 500.
4. So, L.L. and Goldstein, I.J., Biochim. Biophys. Acta,
       (1968), 165, 398.
5. Goldstein, I.J., Cifonelli, J.A. and Duke, J., Biochemistry,
       (1974), 13, 867.
6. Koenig, S.H., Brown, R.D. III and Brewer, C.F., Proc. Nat.
       Acad. Sci., (1973), 70, 475.
7. Yariv, J., Kalb, A.J. and Levitzki, A., Biochim. Biophys.
       Acta, (1968), 165, 303.
8. Brown, R.D., Brewer, C.F. and Koenig, S.H., Biochemistry,
       (1977), 16, 3883.
9. Koenig, S.H. and Schillinger, W.E., J. Biol. Chem., (1968),
       244, 3283.
10. Hallenga, K. and Koenig, S.H., Biochemistry, (1976),
       15, 4255.
11. Steinhardt, J. and Reynolds, J., "Multiple Equilibria in
       Proteins", Academic Press, New York, (1969).
12. McKenzie, G.H. and Sawyer, W.H., J. Biol. Chem., (1973),
       248, 549.
13. Becker, J.W., Reeke, G.N.Jr., Wang, J.L., Cunningham, B.A.
       and Edelman, G.M., J. Biol. Chem., (1975), 250, 1513.

14. Loontiens, F.G., van Wauve, J.P. and DeBruyne, C.K.,
    Carbohydrate Res., (1975), 44, 150.
15. Mildvan, A. and Cohn, M., Advan. Enzymol., (1970), 33, 1.
16. Quiocho, F.A., Bethge, P.H., Lipscomb, W.N., Studebaker,
    J.F., Brown, R.D. and Koenig, S.H., Cold Springs Harbor
    Symposia on Quantitative Biology XXXVI, 561, (1971).
17. Pflumm, M.N., Wang, J.L. and Edelman, G.M., J. Biol. Chem.,
    (1971), 246, 4369.
18. Brewer, C.F., Sternlicht, H., Marcus, D.M. and Grollman,
    A.P., "Lysozyme", Academic Press, New York, (1974).
19. Becker, J.W., Reeke, G.N. Jr., Cunningham, B.A. and Edelman,
    G.M., Fed. Proc., (1976b), 35, 1716.
20. Brewer, C.F., Sternlicht, H., Marcus, D.M. and Grollman,
    A.P., Abstracts, 164th American Chemical Society
    National Meeting, paper # 241, (1972).
21. Brewer, C.F., Sternlicht, H., Marcus, D.M. and Grollman,
    A.P., Proc. Nat. Acad. Sci. USA, (1973a), 70, 1007.
22. Brewer, C.F., Sternlicht, H., Marcus, D.M. and Grollman,
    A.P., Biochemistry, (1973b), 12, 4448.
23. Villafrance, J.J. and Viola, R.E., Arch. Biochem. Biophys.,
    (1974), 165, 51.
24. Alter, G.M. and Magnuson, J.A., Biochemistry, (1974),
    13, 4038.
25. Edelman, G.M., Cunningham, B.A., Reeke, G.N. Jr., Becker,
    J.W., Waxdal, M.J. and Wang, J.L., Proc. Nat. Acad.
    Sci. USA, (1972), 69, 2580.
26. Hardman, K.D. and Ainsworth, C.F., Biochemistry, (1976),
    15, 1120.
27. Becker, J.W., Reeke, G.N. Jr., Cunningham, B.A. and Edelman,
    G.M., Nature, (1976a), 259, 406.
28. Hehre, E.J., Bull. Soc. Chim. Biol., (1960), 42, 1581.
29. Young, N.M. and Leon, M.A., Biochim, Biophys. Acta, (1974),
    365, 418.
30. Kornfeld, R. and Ferris, C., J. Biol. Chem., (1975),
    250, 2614.
31. Teichberg, V.I. and Shinitzky, M., J. Mol. Biol., (1973),
    74, 519.
32. Privat, J.R., Domotte, F., Mialonier, G., Bouchard, P. and
    Monsigny, M., Eur. J. Biochem., (1974), 47, 5.
33. Hardman, K.D. and Ainsworth, C.F., Biochemistry, (1973),
    12, 4442.
34. Banerjee, S.K. and Rupley, J.A., J. Biol. Chem., (1973),
    248, 2117.
35. Allan, A.K., Neuberger, A. and Sharon, N., Biochem. J.,
    (1973), 131, 155.
36. Abdullah, M., French, D. and Robyt, J.F., Arch. Biochem.
    Biophys., (1966), 114, 595.
37. Thoma, J.A., Spradlin, J.E. and Dygert, S., "The Enzymes",
    p. 115, D. Boyer (Ed.), (1971).

RECEIVED September 8, 1978.

4

Binding of p-Nitrophenyl 2-O-α-D-Mannopyranosyl-α-D-Mannopyranoside to Concanavalin A

TAFFY WILLIAMS, JULES SHAFER, and IRWIN GOLDSTEIN

Department of Biological Chemistry, The University of Michigan, Ann Arbor, MI 48109

For many years this laboratory has been studying the interactions of concanavalin A (con A), a carbohydrate binding protein of considerable interest (1-4), with simple and complex carbohydrates (5-8). It was established that compared to monosaccharides, oligosaccharides composed of α-(1→2)-linked D-mannose units exhibited an enhanced affinity for con A (9). This increased affinity for con A was explained both in terms of an extended binding site (5,10) and a statistical mode (11).

In order to further examine the nature of the binding phenomenon, kinetic and thermodynamic parameters of the binding of p-nitrophenyl 2-O-α-D-mannopyranosyl-α-D-mannopyranoside (M_2) (see Figure 1) were compared to p-nitrophenyl α-D-mannopyranoside (M_1) (and its 2-O-methyl derivative). The above disaccharide contains two α-D-mannopyranosyl residues both of which are capable of interacting with con A. Any enhanced affinity of M_2 toward con A due to a statistical effect should be reflected by differences in the entropy of activation for binding M_2 and the corresponding monosaccharide. No difference should be seen in the enthalpy of activation for binding the mono- and disaccharide since only entropy terms are affected in true statistical modes of binding. Therefore, the activation parameters for the binding of M_2 to con A were examined in order to determine whether con A interacts with both mannopyranosyl residues at an extended binding site or the enhanced affinity of M_2 for con A is due to a statistical effect whereby the two mannopyranosyl residues of M_2 simply increase the probability of forming a complex with the protein.

Initial observations indicated that con A as isolated by the conventional Sephadex procedure (9) exhibited heterogeneity with respect to binding M_2. Spectral studies of the interactions of con A with M_2 indicated that con A contains at least two components which interact differently with M_2. The simple reaction,

$$P+L \rightleftharpoons PL, \qquad (1)$$

represents the formation of a con A-ligand complex wherein the

concentration of PL is determined by the concentrations of P and
L. The assignment of the terms P and L to ligand and protomer is
arbitrary, and symmetrically interchanging the concentrations of
protein and ligand should not affect the concentration of the complex. Such behavior is observed when one studies the interaction
of con A with monosaccharides (7,8). However, when M_2 is mixed
with con A prepared by the conventional Sephadex procedure the
amount of complex formed appears to depend upon which ligand is in
excess. Figure 2 depicts the asymmetric response observed by
means of difference spectra for formation of a con A-M_2 complex
when solutions of con A and M_2 are mixed. It is evident that the
apparent amount of complex formed as reflected by the magnitude of
the difference spectra, depends on which component is in excess.
Such a result could be explained if M_2 were impure, or if the con A
were heterogeneous with respect to its interactions with M_2. M_2
was examined by several methods and found to be pure. To characterize the heterogeneity of the con A-M_2 interaction, equilibrium
dialysis studies were performed with the concentrations of M_2
equal to twice the concentration of binding sites. The concentration of both protein and ligand were kept high to ensure occupancy
of all the binding sites. A plot of M_2 bound versus the concentration of binding sites (Figure 3) gave a straight line with a
slope of 0.9 demonstrating that nearly all the protein combining
sites are capable of binding M_2. However, all of these con A
binding sites do not necessarily have the same affinity for M_2 as
indicated by the nonlinear Scatchard plot in Figure 4. In this
figure, the line drawn through the points is a computer fit of the
data for a two component system in which M_2 binds each component
with a different affinity.

   Since con A as isolated conventionally (9) exists as a mixture of intact and nicked protomeric subunits (12), a method was
developed to purify large quantities of these species so that they
might be examined as a source of the heterogeneous-like behavior
seen in the Scatchard plot (Figure 4) and ultraviolet difference
spectra (Figure 2). Con A was applied to a Sephadex G-75 column
equilibrated at pH 7.2 (0.02 M Tris-HCl) and 4°C. Elution with 19
mM D-glucose, in 0.02 M Tris-HCl (pH 7.2) resulted in the displacement of a fraction of protein composed of 70-80% nicked subunits. Further elution with 100 mM D-glucose gave a second peak
composed of 100% intact subunits. These results are indicated in
Figure 5. In this procedure, about 40% of the protein remained
on the column.

   A Scatchard plot obtained for the interaction of M_2 and
"intact" con A, Figure 6, was found to be linear as would be expected for single component systems and a value of 1.9×10^5 M^{-1}
was obtained for the association constant. Also, identical difference spectra were obtained when the concentration of ligand and
protein were symmetrically interchanged. Both of these results
demonstrate the homogeneous behavior of the "intact" con A with
respect to M_2. Interestingly, the "nicked" con A that was obtained

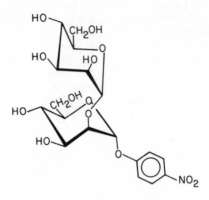

p -Nitrophenyl
2 - *O* - α- D -Mannopyranosyl - α- D - Mannopyranoside

Figure 1. Structure of p-*nitrophenyl* 2-*O*-α-D-*mannopyranosyl*-α-D-*mannopyrano*-*side*

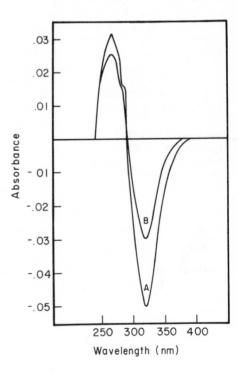

Figure 2. Difference spectra observed on mixing M_2 and con A obtained commercially from Calbiochem or prepared according to Ref. 9. A, 40 μM M_2 and 100 μM con A; B, 100 μM M_2 and 40 μM con A.

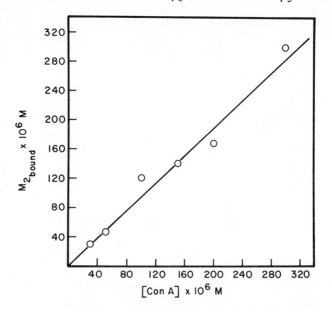

Figure 3. The dependence of bound ligand ($[M_2]_b$) on the total concentration of protomeric units ($[P]_t$) of con A from Calbiochem. The total concentration of M_2 was maintained at twice $[P]_t$.

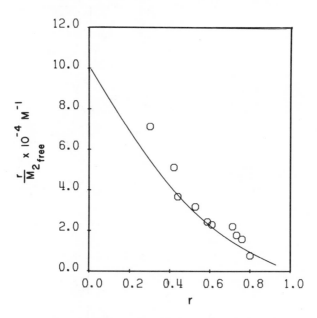

Figure 4. Scatchard plot for the binding of M_2 to commercially prepared con A as determined by equilibrium dialysis. The total concentration of con A protomeric units was 15 μM. The solid line represents a computer simulated fit of the data for a two component system.

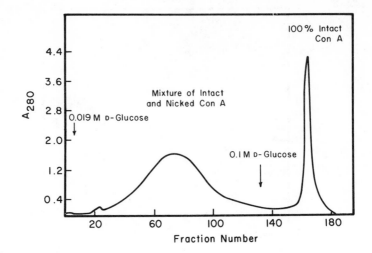

Figure 5.   Elution profile obtained for Calbiochem con A on Sephadex G-75 in a Tris-HCl pH 7.2 buffer. Elution was effected with 0.019 M glucose followed by 0.1 M glucose.

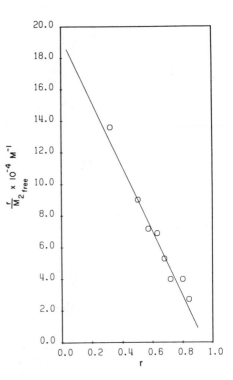

Figure 6.   Scatchard plot for the binding of M_2 to intact con A as determined spectrophotometrically

from the Sephadex G-75 chromatography was indistinguishable from the intact con A in its interactions with M_2. Furthermore, mixtures of intact and nicked con A behaved as a single homogeneous protein with respect to M_2. Therefore, we conclude that the material which remains bound to the Sephadex G-75 column gives rise to the heterogeneous response to M_2. The elution and characterization of the protein fraction which remains bound on the column is currently under investigation.

The fraction of protein, eluted from Sephadex G-75 which behaves as a single homogeneous protein with respect to its interactions with M_2 was also studied with regard to its interaction with monosaccharides. The association constant of the interaction of M_2 with intact con A is 19 times greater than that obtained for the interaction of p-nitrophenyl α-D-mannopyranoside with con A. The normalized spectra (Figure 7) of the monosaccharide and disaccharide complexes were found to be very similar suggesting that the interactions between con A and the nitrophenyl group which causes the spectral perturbation are the same for both complexes.

The "intact" con A which behaves homogeneously toward M_2 was used in further kinetic studies. Stopped-flow studies were performed on a structural analog of M_2, p-nitrophenyl 2-O-methyl-α-D-mannopyranoside (Figure 8). In these experiments, con A protomer concentrations were 10 times greater than ligand concentrations and pseudo-first order kinetics were observed. Complex formation between p-nitrophenyl 2-O-methyl-α-D-mannopyranoside was found to be a monophasic process. The reaction between con A and this M_2 analog was similar to that observed for con A and other monosaccharides (7). As described previously (7), a plot of the observed rate constants with respect to the concentration of ligand in excess should be a straight line which intersects the Y-axis at a point equal to the dissociation rate constant i.e. the plot should follow the equation $k_{obs} = k_1[A]+k_{-1}$ wherein A is the component in excess. Such a plot of the M_2 analog demonstrates this pseudo-first order behavior (Figure 9). The off-rate was determined by displacement of the chromogenic monosaccharide from con A with methyl α-D-mannopyranoside.

Although binding of monosaccharides and intact con A proceed by simple monophasic pseudo-first order kinetics, binding of M_2 does not. As can be seen in Figure 10, binding is a biphasic process consisting of a faster initial phase and a slower later phase. Such a biphasic profile is obtained regardless of the reagent in excess and is unaffected by the presence or absence of metals. On further examination the initial portion of the reaction was found to be concentration dependent and the later portion of the reaction was found to be concentration independent. The dissociation was found to be monophasic and first-order.

Of the possible pathways for binding, the model,

$$\text{PL*} \underset{k_{21}}{\overset{k_{12}}{\rightleftharpoons}} \text{P+L} \underset{k_{32}}{\overset{k_{23}}{\rightleftharpoons}} \text{PL}$$

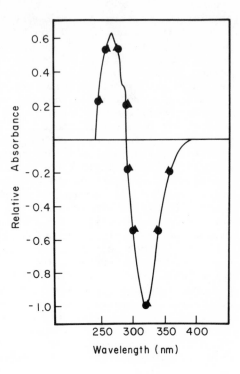

Figure 7. Normalized difference spectra for con A-M₂ and con A-p-nitrophenyl α-D-mannopyranoside complexes. The relative absorbance is defined as the ratio of the absorbance difference at a given wavelength to the absolute value of the absorbance difference at 317 nm. Spectra were obtained by mixing 200 μM "intact" con A and 40 μM ligand: (○) con A-M₂; (△) con A-p-nitrophenyl α-D-mannopyranoside.

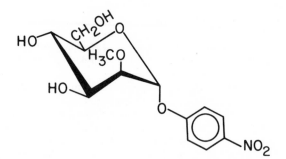

p -Nitrophenyl
2-*O*-Methyl-α-D-Mannopyranoside

Figure 8.   Structure of p-nitrophenyl 2-O-methyl-α-D-mannopyranoside

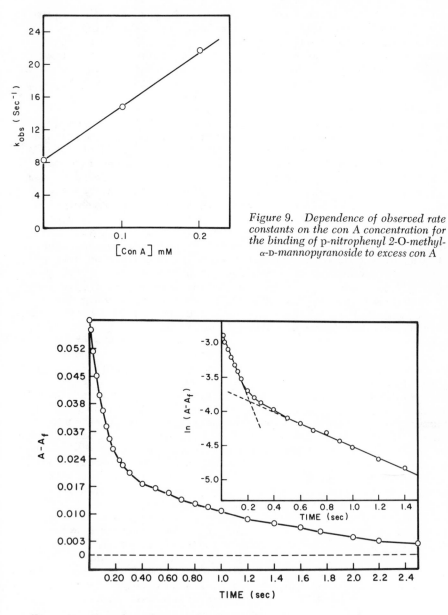

Figure 9.   *Dependence of observed rate constants on the con A concentration for the binding of p-nitrophenyl 2-O-methyl-α-D-mannopyranoside to excess con A*

Figure 10.   *Time dependence of absorbance obtained after mixing con A (200 μM final concentration) and M₂ (20 μM final concentration)*

wherein k_{12} = 1.1 sec^{-1}, k_{21} = 2.3 x 10^4 sec^{-1} $\underline{\underline{M}}^{-1}$, k_{32} = 0.22 sec^{-1}, k_{23} = 3.2 x 10^4 sec^{-1} $\underline{\underline{M}}^{-1}$, $\Delta\varepsilon^*_{PL}$ = 0, and $\Delta\varepsilon_{PL}$ = -3516 cm^{-1} $\underline{\underline{M}}^{-1}$, best fits the data. In this model, protein (P) and ligand (L) combine to initially form two different complexes at relative rates equal to the ratio k_{21}/k_{23}. This is followed by a slow phase in which PL and PL* equilibrate and the concentration of PL* relative to that of PL approaches the equilibrium ratio $k_{21} k_{32}/ k_{12} k_{23}$.

In order to characterize further the interactions between con A and mono- and di-saccharides, the activation parameters for the binding of M_2 to con A were determined from the temperature dependence of the observed kinetic parameters. Plots of rate constants for the fast and slow phases as well as the rate constant for dissociation of the con A-M_2 complex were found to vary linearly with 1/T (Figure 11). The values of the entropy and enthalpies of activation obtained from the rates of association were found to be different for interactions with ligands containing one and two D-mannopyranosyl residues (Table I). This result implies that a statistical model cannot explain the enhanced binding of the disaccharide since both the enthalpy and entropy terms vary from those obtained for a monosaccharide. Furthermore, at 25°C the association constant for the predominant complex (PL) is 14.5 times greater than that observed for the monosaccharide signifying that in PL there are extended interactions between the carbohydrate and protein. These facts in addition to the different binding kinetics of M_2 compared to monosaccharides are difficult to rationalize in terms of a statistical model. These results are best explained by a model in which interactions occur simultaneously with groups on both mannopyranosyl residues.

Four major conclusions may be reached as a result of these studies. First, con A as prepared conventionally or as obtained commercially behaves as a two component system in which each component interacts differently with M_2. This point was demonstrated by spectral methods, Scatchard analysis of the data and by purification of a fraction of con A which behaves homogeneously with respect to M_2. Secondly, con A interacts simultaneously with groups on both glycosyl moieties in M_2. This conclusion is supported by the large difference in the affinities of con A with M_2 as compared to M_1, the difference in the enthalpies of activation for binding M_2 and monosaccharides, and the different kinetic process seen for M_2 and monosaccharide binding. Thirdly, the interaction between con A and the nitrophenyl group which is responsible for the spectral perturbation appears to be the same in both the con A-M_2 and con A-M_1 complexes as detected by the similar normalized difference spectra of the two complexes. Finally, con A is found to bind carbohydrate ligands in more than one orientation. This conclusion is derived from the reorientation model, wherein two types of con A-ligand complexes form.

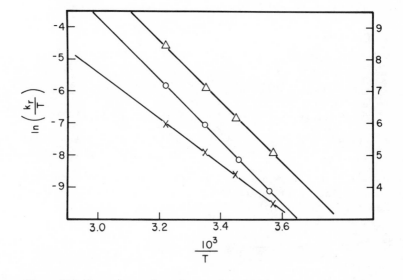

Figure 11.   Temperature dependence of k_{fast} ($\times$), k_{slow} ($\triangle$), *and* k_{off} ($\bigcirc$)

Table I

Activation Parameters for Formation and Breakdown of
Complexes of Con A with M_2 and With
p-Nitrophenyl α-$\underline{D}$-Mannopyranoside

| | Formation | Breakdown | |
	k_{fast}	k_{off}	k_{slow}
Monosaccharide Complex[a]			
$\Delta H^\dagger$ (kcal/mole)		9.5 (±0.3)	16.8 (±0.2)
$\Delta S_u{}^\dagger$ (e.u)		2.8 (±1.1)	1.3 (±0.7)
Disaccharide Complex[b]			
$\Delta H^\dagger$ (kcal/mole)	13.8 (±0.3)	19.4 (±0.1)	19.0 (±0.1)
$\Delta S_u{}^\dagger$ (e.u.)	16.6 (±1.1)	3.3 (±0.3)	4.3 (±0.5)

(a) Values from reference 2.   (b) Activation parameters were determined by interpreting linear plots (Figure 11) of ln (k_r/T) $\underline{vs}$
1/T according to the relationship

$$\ln \frac{k_r}{T} = - \frac{\Delta H^\dagger}{R} \frac{1}{T} + \ln \frac{k}{h}$$

where k_r is the kinetic parameter k_{fast}, k_{off}, or k_{slow} and k/h is
the ratio of the Boltzmann constant to the Planck constant.   The
entropy of activation for association contains the entropy change
for bringing two molecules together in $1\underline{M}$ solution.   Unitary
entopies of activation ($\Delta S_u{}^\dagger$) do not contain this entropy of mixing and were estimated using the relationship (13)

$$\Delta S_u{}^\dagger = \Delta S^\dagger + 7.98$$

For dissociation there is no change in the number of molecules on
formation of the activated complex and $\Delta S_u{}^\dagger = \Delta S$. $\Delta S_u{}^\dagger$ also was
equated to $\Delta S^\dagger$ for the first-order rate constant k_{slow}.

Literature Cited

1. Goldstein, I. J., and Hayes, C. E., Adv. Carbohyd. Chem. Metabol. (1978) 35, 127-340
2. Lis, H., and Sharon, N., Ann. Rev. Biochem. (1973) 42, 541-574
3. Chowdhury, T. K., and Weiss, A. K. (eds.) Advances in Experimental Medicine and Biology 55, pp. 1-359 (1974)
4. Bittinger, H., and Schnebli, H. P. (eds.) Concanavalin A as a Tool, pp. 1-639, John Wiley and Sons, New York (1976)
5. Goldstein, I. J., Reichert, C. M., and Misaki, A., Ann. of N.Y. Acad. Sci. (1974) 234, 283-296
6. Bessler, W., Shafer, J. A., and Goldstein, I. J., J. Biol. Chem. (1974) 249, 2819-2822
7. Lewis, D. S., Shafer, J. A., and Goldstein, I. J., Arch. Biochem. Biophys. (1976) 172, 689-695
8. Hassings, G. S., and Goldstein, I. J., Eur. J. Biochem. (1970) 16, 549-556
9. Agrawal, B. B. L., and Goldstein, I. J., Biochim. Biophys. Acta (1967) 147, 262-271
10. So, L. L., and Goldstein, I. J., J. Biol. Chem. (1968) 243, 2003-2007
11. Brown, R. D., Brewer, C. F., and Koenig, S. H., Advances in Experimental Medicine and Biology (1974) 55, 323-324
12. Wang, J. L., Cunningham, B. A., and Edelman, G. M., Proc. Natl. Acad. Sci. USA (1971) 68, 1130-1134
13. Kauzmann,W., in "Advances in Protein Chemistry" (Anfinsen, C. B. Jr., Anson, M. L., Bailey, K., and Edsall, J. T., eds.), Vol. 14, pp. 33-35, Academic Press, New York

RECEIVED September 8, 1978.

5

The Use of Lectins to Study Cell Surface Glycoconjugates

R. D. PORETZ

Biochemistry Department, Bureau of Biological Research, Rutgers University,
New Brunswick, NJ 08903

In recent years, as interest in the properties of biological
membranes has increased, researchers have used lectins to study
the role of the binding of multivalent ligands to the plasms membrane in the induction of specific biological responses. This
report will detail the use of lectins to study the structure of
the carbohydrate moiety of lymphocyte surface bound H-2 glycoproteins. Furthermore, it will describe a novel lectin-induced perturbation of the metabolism of in vitro cultured fibroblasts which
results in the massive accumulation of lysosomes in these cells.
This response is greatly reduced in cells which have been transformed by oncogenic viruses.

H-2D Antigens

The H-2K and H-2D glycoproteins are two members of a group of
membrane antigens which are gene products of the H-2 histocompatibility complex of the mouse. The serological properties of these
histocompatibility antigens have classically been determined by
developing specific alloantisera which can detect differences in
allelic forms of these antigens (1).
In recent years the biochemical properties of the H-2K and
H-2D histocompatibility antigens have been studied primarily by
Nathenson (cf. ref. 2) and Davies (3), and more recently by Edelman (4) and Hood (5). It is now apparent that both antigens represent related structures composed of a long polypeptide chain of
approximately 45,000 daltons containing one or more carbohydrate
chains and a small polypeptide of 11,600 daltons devoid of carbohydrate (cf. ref. 6). The small polypeptide chain which is identical
to the β_2-microglobulin, possesses a structure which is apparently
common to H-2K and H-2D antigens of different histocompatibility
types (7) and the large glycoprotein chain appears to possess the
structural differences defining the reactions of these antigens
with various alloantisera (8). Though a significant portion of
the large chain of the H-2K and D antigens is composed of carbohydrate, little is known about the structure of the carbohydrate

0-8412-0466-7/79/47-088-056$05.00/0

moiety of the antigen.   It represents approximately 7% of the
antigen by weight and is composed of sialic acid, $\underline{N}$-acetyl-$\underline{D}$-
glucosamine, $\underline{D}$-mannose, $\underline{D}$-galactose and $\underline{L}$-fucose, presumably in
a serum-type linkage ($\underline{9}$).   Nathenson and Cullen ($\underline{2}$) have suggested
that the carbohydrate chain does not possess structural features
which are capable of reacting with complement mediated cytotoxic
alloantisera.   However, the contribution of the structure of the
carbohydrate chain to the physiological function of the antigen,
that is, its involvement in the induction of a cellular immune
response and the ability of this moiety to induce non-cytotoxic
antibody is unknown.

   We have approached the problem of studying the structure of a
specific cell surface carbohydrate moiety on a complex cell surface by determining the ability of carbohydrate binding reagents
(lectins) to react with the specific carbohydrate structure under
investigation ($\underline{10}$).   The glycoprotein of concern is selected from
the mass of other surface glycoconjugates by use of monospecific
antisera directed toward the polypeptide portion of the glycoprotein.   This approach has allowed us to determine structural features of the carbohydrate moiety of the cell surface bound-$\underline{H-2D}$
glycoprotein antigen.   Employing monospecific anti-$\underline{H-2.4}$ serum
which is directed against the polypeptide portion of the antigen
($\underline{8}$) and various $\underline{N}$-acetyl-$\underline{D}$-galactosamine and $\underline{D}$-galactose binding
lectins from $\underline{Sophora}$ $\underline{japonica}$ ($\underline{11}$), $\underline{Maclura}$ $\underline{pomifera}$ ($\underline{12}$), and
$\underline{Wistaria}$ $\underline{floribunda}$ ($\underline{13,14}$), and the $\underline{D}$-mannose binding lectin,
concanavalin A ($\underline{15}$), we have studied the competitive binding ac-
tivities of these carbohydrate reagents and anti-$\underline{H-2.4}$ immunoglobulin in leucoagglutination, cytotoxicity and complement binding
assays, and cellular binding of fluorescent and radiolabelled proteins.   In addition, we have examined the effect of these reagents
on the induction of lymphocyte receptor mobility.

   The following experiments will demonstrate that the use of
such techniques is both possible and practical to study the $\underline{H-2D}$
antigens.   The ability of lectins to inhibit the specific binding
of anti-$\underline{H-2.4}$ IgG to lymphocytes was detected by the complement
depletion technique.   This assay measures the amount of exogenous
complement remaining subsequent to the formation of IgG-lymphocyte
complexes which are capable of fixing complement.   The residual
complement activity, as measured by complement mediated lysis of
hemolysin treated sheep erythrocytes is a measure of the binding
of anti-$\underline{H-2D}$ antibody to lymphocytes.   Figure 1 demonstrates that
three out of the five lectins tested significantly inhibit, in a
dose related fashion to a level of 88-90%, the ability of the
lymphocyte-IgG reaction mixture to fix complement.   The most po-
tent lectin tested is the hemagglutinin of $\underline{M}$. $\underline{pomifera}$ seeds,
requiring only 2.1 µg/ml to cause 50% inhibition of potential complement fixation by the IgG.   This is one-fifth the amount of $\underline{W}$.
$\underline{floribunda}$ mitogen and one-twentieth the amount of concanavalin A
needed to cause an equivalent level of inhibition.   However, 100%
inhibition of complement fixation was not attained even at concen-

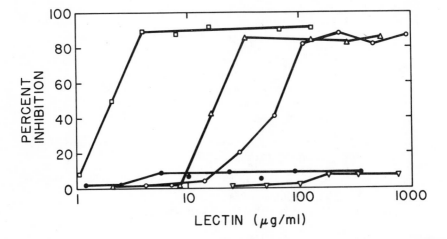

*Figure 1. Inhibition by lectins of complement fixation by the anti-*H-2.4-*lympho-cyte reaction. (□) M. pomifera agglutinin; (△) W. floribunda mitogen; (○) con-canavalin A; (●) W. floribunda agglutinin; (▽) S. japonica agglutinin.*

trations of lectin 10-fold greater than the minimum amount needed
to cause maximum inhibition (88-90%). The hemagglutinins of W.
floribunda and S. japonica appear to cause only a small degree of
inhibition of the binding of anti-H-2 IgG to lymphocytes.

In order to determine if lectins affect the reactions of complement mediated hemolysis subsequent to binding of antibody to
the lymphocyte we have employed a synthetic particulate antigen.
This was accomplished by testing the ability of lectin to inhibit
the fixation of complement by the complex of anti-human serum
albumin (HSA) and HSA-conjugated aminoethyl Biogel beads. The
HSA-conjugated aminoethyl Biogel beads may be considered to be
cell-like particles coated with a carbohydrate-free protein.
Thus, the anti-HSA immunoglobulin will be able to react with these
particles, but lectins should be unreactive and not competitively
interfere with the binding of antibody to the beads. Though the
antibody-conjugated bead complex is capable of fixing complement,
all the lectins utilized in this work cause no more than a 6%
decrease in the fixation of complement, even at concentrations in
excess of the amount of lectin needed to cause maximum inhibition
of complement in the anti-H-2.4-lymphocyte reaction.

To show that the H-2D antigen possesses the lectin binding
structures implied from the results of the inhibition of complement depletion experiment we have employed a co-capping technique.
This approach can demonstrate the coincident mobility within the
lymphocyte membrane, of structures capable of binding H-2 antibody
and lectins. The assay is conducted by first allowing anti-H-2
antibody to induce the accumulation of H-2D antigen within one
area of the cell surface (forming a cap), and then, under conditions restraining further mobility of membrane receptors, to
determine the location of specific lectin binding structures by
use of immunofluorescence. Table I demonstrates that those lectins which caused significant inhibition of complement fixation by
anti-H-2.4 IgG appear to co-cap with the H-2 antigen. That is,
the number of cells which have anti-H-2.4 IgG induced caps as visualized with fluoresceinated anti-H-2.4 IgG, correlated well with
the number of cells which have anti-H-2.4 induced lectin binding
receptors [as visualized with a double sandwich of rabbit anti-
lectin serum and fluoresceinated goat anti-rabbit gamma globulin
(F-GARG)]. However, those lectins which cause little inhibition
of binding of anti-H-2.4 IgG, as detected by the complement
depletion assay, do not co-cap with the H-2.4 antigen. Consistent
with these results, inhibitory lectins also cause a decrease in
the intensity of fluorescence of lymphocytes which have been reacted with fluorescein labelled anti-H-2.4 IgG (Table II). Those
lectins which do not affect the binding of antibody to the lymphocyte or co-cap with the H-2D antigen had no affect on the level of
fluorescence of the anti-H-2.4 IgG labelled cells. It is noteworthy that preincubation of cells with unlabelled anti-H-2.4 IgG
is unable to alter immunofluorescence of the cells labelled with
the lectins used in this study (lectin; rabbit anti-lectin serum;

F-GARG) suggesting the presence of non-H-2D lectin receptors on these lymphocytes.

TABLE I. Co-capping of Lectin Receptors and H-2d Structure

First Treatment*	Second Treatment[+]	% Labelled Cells Capped
FITC[†] Anti-H-2D IgG	No lectin nor anti-lectin	51
Anti-H-2D IgG	W. floribunda mitogen	45
Anti-H-2D IgG	M. pomifera lectin	34
Anti-H-2D IgG	Concanvalin A	45
Anti-H-2D IgG	W. floribunda agglutinin	2
Anti-H-2D IgG	S. japonica lectin	<1

*Antiserum at 37°, 15 min; Wash 0°.
[+]Lectin 100-300 µg/ml at 0°, 30 min; Wash, 0°; Rabbit anti-lectin serum, 0°, 15 min; Wash 0°; Fluorescein labelled goat anti-rabbit IgG, 0°, 15 min; Wash 0°; View.
[†]FITC, Fluorescein isothiocyanate.

TABLE II. Inhibition of Immunofluorescence

First Treatment*	Second Treatment[+]	Relative Intensity of Fluorescence
None	FITC[†] Anti-H-2	100%
W. floribunda mitogen	FITC Anti-H-2	50%
M. pomifera lectin	FITC Anti-H-2	25%
Concanavalin A	FITC Anti-H-2	25%
S. japonica lectin	FITC Anti-H-2	95%
W. floribunda agglutinin	FITC Anti-H-2	95%

*Lectin 100-200 µg/ml at 0°, 15 min; Wash 0°.
[+]Fluorescein labelled-anti-H-2.4 IgG at 0°, 15 min; Wash 0°; View.
[†]FITC, Fluorescein isothiocyanate.

Additional experiments were designed to measure directly the ability of lectins to competitively inhibit the binding to lympho-cytes of radiolabelled F(ab)' fragments of anti-H-2.4 IgG. Figure 2 shows the concentration dependence of inhibition by lectin of the binding of ^{125}I F(ab)' fragments to lymphocytes. It is appar-ent that those lectins which inhibit the binding of anti-H-2.4 IgG to lymphocytes as measured by the complement depletion assay also cause inhibition of binding of the labelled F(ab)' fragments to those cells. Sophora japonica which causes only slight inhibition of complement fixation by the anti-H-2.4 IgG has only a minor effect on the binding of the F(ab)' fragments to lymphocytes. It is further evident that each inhibition of binding curve exhibits at least two separate regions of positive slope. There is a 50-

100-fold difference in concentration of an inhibitory lectin, required to reach the two points of inflection of the positive slopes on each curve. Such curves, which are indicative of an apparent heterogeneity of binding, are not due to the inherent heterogeneous nature of the IgG as shown in Figure 3. It may be seen that, though, the binding of F(ab)' fragments to lymphocytes develop a scatchard plot indicating heterogeneity of binding, this apparent heterogeneity is far from that expected for two major populations of fragments with a 100-fold difference in binding affinities. It is also noteworthy that we were unable to cause greater than 42% inhibition of binding of the F(ab)' fragment by the W. floribunda mitogen or 60% inhibition by the M. pomifera lectin even at concentrations of 400 µg/ml of the lectins.

The data presented here demonstrate that lectins may be used to explore structural aspects of the carbohydrate moieties of specific cell surface bound antigens. Furthermore, it is suggested that the H-2D antigen possesses differences in the structure of the carbohydrate moieties which may reflect microheterogeneity due to the biosynthesis of only partial structures, differences in the number of sialic acid residues, the relationship of the carbohydrate chains to the specific antigenic determinants and/or major structural differences of the carbohydrate chains on different H-2D antigens.

Lectin Induced Accumulation of Lysosomes in Fibroblasts

Responses of cells to membrane reactive substances in both physiological and non-physiological situations are well-documented. In recent years studies on the in vitro interaction of lectins with animal cells have demonstrated that the binding of such multivalent carbohydrate binding proteins to the cell surface may induce responses evident in changes in the plasma membrane and intracellular metabolism of the cells. Lectins have been reported to affect lymphocytes in a manner resulting in: the controlled movement of plasma membrane glycoproteins (cf. ref. 16); changes in ion flow (17); increased phospholipid turnover in plasma membranes (18); and induction of mitosis (19). Similarly, lectins have been shown to affect non-lymphoid cells. These properties include: cytotoxicity (20); inhibition of amino acid transport (21); activation of hexose transport (22); and inhibition of DNA synthesis (23).

Recently, in collaboration with R. J. Kuchler and D. Cryan of the Bureau of Biological Research, Rutgers University, we have observed that Balb/c 3T3 fibroblasts when cultured for prolonged periods with 25-400 µg/ml of purified W. floribunda agglutinin accumulated massive amounts of vacuoles. This treatment with lectin is apparently not toxic to the cells as measured by the ability of treated cells to incorporate tritiated thymidine into cellular DNA. Figure 4 shows the effect on cells of treatment for 48 hours with 150 µg/ml of lectin as compared to cells treated in

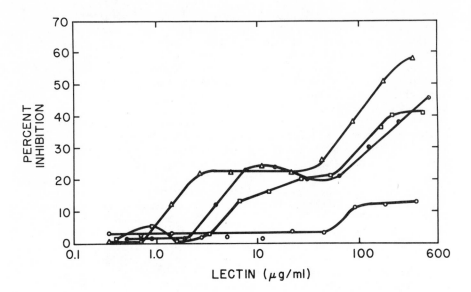

Figure 2. Inhibition by lectins of the binding of ^{125}I F(ab)′ fragments of anti-H-2.4 IgG to lymphocytes. (△) M. pomifera agglutinin; (⊙) concanavalin A; (□) W. floribunda mitogen (○) S. japonica agglutinin.

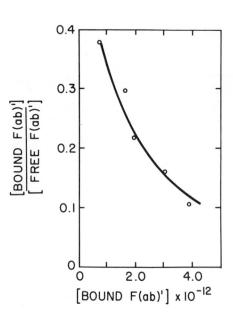

Figure 3. Scatchard plot of the binding of ^{125}I F(ab)′ fragments of anti-H-2.4 IgG to lymphocytes

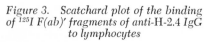

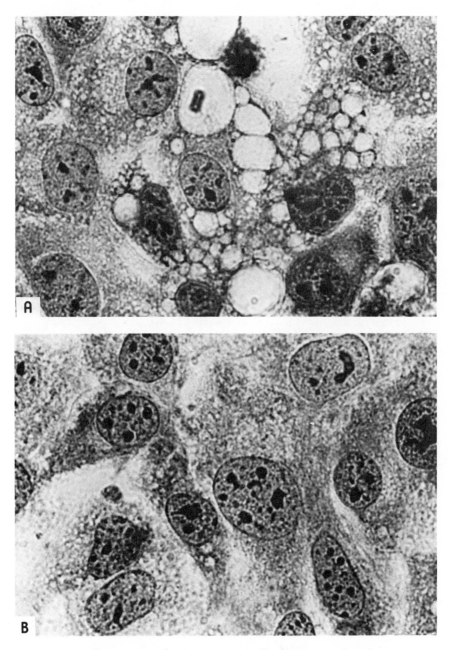

Figure 4. Photomicrograph of 3T3 murine fibroblasts stained with Wright's-Giemsa stain: a) treated with W. floribunda agglutinin (150 µg/ml) in Eagle's Minimum Essential Medium with Hanks' salts for 48 hr; b) control

a like manner but without lectin. The phenomenon appears more dramatic when cells are vitally stained with acridine orange under conditions which cause concentration of this fluorescent dye in lysosomes (24). It is evident from Figure 5, which shows such stained cells as viewed by dark-field fluorescence microscopy, that the vacuoles induced by the lectin concentrated the acridine orange in a manner characteristic of lysosomes (24). Though not apparent here, the lysosomes fluoresce bright red when excited with light at 490 nm. The induction of the lysosomes by W. floribunda hemagglutinin is specific in that the lectins from M. pomifera, S. japonica, Canavalia ensiformis and wheat germ have no effect on the cells even at concentrations of 400 µg/ml for up to 48 hours. Furthermore, the ability of W. floribunda lectin to induce the accumulation of lysosomes in these cells is completely abrogated when lactose (a known inhibitor of the hemagglutinating activity of the lectin) is added to the culture simultaneously with the lectin. Interestingly we have also noted that four-five times greater concentrations of lectin are required to affect transformed murine fibroblasts (MSV - transformants of 3T3 cells and L-cells) than needed to cause an equivalent response with 3T3 cells.

Recently, Lotan et al. (25) have described the lectin induced vacuolation of macrophages, cells normally highly phagocytic. These authors suggest that the multivalency of the lectins is an important factor in vacuole formation in these cells. However, we have seen that two of those lectins most active in the macrophage system, namely concanavalin A and wheat germ agglutinin, are unable to induce lysosomes in cultured fibroblasts. We are presently studying the properties of the lectins and the cells in order to more completely understand the mechanism of lectin induced accumulation of lysosomes in animal cells.

Acknowledgements

This research was supported by grants from the National Cancer Institute (CA-20889 and CA-17193).

References

1. Shreffler, D. C. and David, C. S., Adv. Immunol. (1975) 20: 125.
2. Nathenson, S. G. and Cullen, S., Biochim. Biophys. Acta (1974) 344:1.
3. Hess, M. and Davies, D. A. L., Eur. J. Biochem. (1974) 41:1.
4. Henning, R., Milner, R., Reske, K., Cunningham, B. A. and Edelman, G. M., Proc. Natl. Acad. Sci (USA) (1976) 73:118.
5. Silver, J. and Hood, L., Proc. Natl. Acad. Sci. (USA) (1976) 73:599.
6. Snell, G. A., Dausset, J. and Nathenson, S., "Histocompatibility," p. 295, Academic Press, New York, 1976.

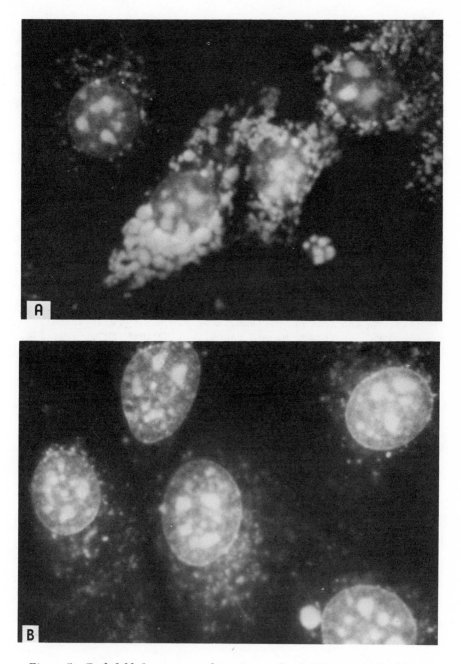

Figure 5. Dark-field fluorescence photomicrograph of 3T3 murine fibroblasts vitally stained with acridine orange: a) treated with W. floribunda agglutinin (75 μg/ml) in Eagle's Minimum Essential Medium with Hanks' salts for 48 hr; b) control

7. Natori, T., Katagiri, M., Tanigaki, N. and Pressman, D.,
    Transplantation (1974) 18:550.
8. Pancake, S. J. and Nathenson, S. G., J. Immunol. (1973) 11:
    1086.
9. Muramatsu, T. and Nathenson, S. G., Biochemistry (1970) 9:
    4875.
10. Janeczek, W. A. and Poretz, R. D., J. Supramol. Struc. (1977)
    Suppl. 1:4.
11. Poretz, R. D., Riss, H., Timberlake, J. W. and Chien, S-M.,
    Biochemistry (1974) 13:250.
12. Bausch, J. N. and Poretz, R. D., Biochemistry (1977) In press.
13. Cheung, G., Haratz, A., Katar, M. and Poretz, R. D.,
    Abstracts of Papers, Chem. Cong. N. A. Cont. 1:BMPC 19.
14. Kaladas, P. and Poretz, R. D., manuscript in preparation.
15. Poretz, R. D. and Goldstein, J. J., Biochemistry (1970) 9:
    2890.
16. Nicolson, G. L., Biochim. Biophys. Acta (1976) 457:57.
17. Crumpton, M. J., Auger, J., Green, M. N. and Maino, V. C., in
    "Mitogens in Immunology," Oppenheim, J. J. and Rosenstreich,
    D. L. (eds.), p. 85, Academic Press, New York, 1976.
18. Fisher, D. B. and Mueller, G. C., Biochim. Biophys. Acta
    (1971) 248:434.
19. Lis, H. and Sharon, N., Ann. Rev. Biochem. (1973) 42:541.
20. Shoham, J., Inbar, M. and Sachs, L., Nature (1970) 227:1244.
21. Inbar, M., Ben-Bassat, H. and Sachs, L., J. Membr. Biol.
    (1971) 6:195.
22. Czech, M. P., Lawrence, J. C. and Lynn, W. S., J. Biol. Chem.
    (1974) 249:7499.
23. Stegman, S. J., Bonfilio, N. D., Fukuyama, K. and Epstein, W.
    L., Cell Differentiation (1974) 3:71.
24. Allison, A. C. and Young, M. R., in "Lysosomes in Biology and
    Pathology," Vol. 2, Dingle, J. T. and Fell, H. B. (eds),
    p. 600, American Elsevier Publishing Co., New York, 1969.
25. Lotan, R., Sharon, N. and Goldman, R., in "Progress in Clini-
    cal and Biological Research," Revel, J. P., Henning, U. and
    Fox, C. F. (eds.), Vol. 17, p. 531, Alan R. Liss, Inc., New
    York, 1977.

RECEIVED September 8, 1978.

Circular Dichroism and Saccharide-Induced Conformational Transitions of Soybean Agglutinin

MICHAEL W. THOMAS, JEANNE E. RUDZKI, EARL F. WALBORG, JR., and BRUNO JIRGENSONS[1]

The University of Texas Cancer Research Center, Department of Biochemistry, M. D. Anderson Hospital and Tumor Institute, Houston, TX 77030

Lectins, sugar-binding proteins, have become powerful molecular probes to investigate the structure, topography and dynamics of cell-surface saccharide determinants (1). The utility of these proteins in the study of the surface properties of a variety of cell types has stimulated renewed interest in the determination of the molecular basis of their saccharide specificity. Furthermore lectins provide relatively simple models for the investigation of noncovalent interactions between saccharides and proteins.

Since the discovery of soybean agglutinin (SBA) by Liener in soybean (Glycine max) extracts (2), a number of papers (3-5) have been published on its isolation, characterization, and sugar-binding specificity. SBA is a glycoprotein comprised of four peptide subunits of approximately 30,000 daltons each (4). Two different types of subunits have been demonstrated by electrophoresis in the presence of sodium dodecyl sulfate (6). This lectin interacts specifically with 2-acetamido-2-deoxy-D-galactose (GalNAc) and galactose (Gal) (4) and possesses two saccharide binding sites per 120,000 daltons (5).

Circular dichroism (CD) has been utilized to investigate the effect of saccharides on the conformation of lectins in solution. CD has demonstrated saccharide-induced conformational changes in the lectins from Canavalia ensiformis (7), Dolichos biflorus (8), Ricinus communis (9), and Triticum vulgaris (10). The present study uses circular dichroism to assess the secondary structure of SBA and to measure conformational transitions induced by the saccharides which bind to this lectin. These and previous studies will contribute to a clearer understanding of the unique properties of these sugar-binding proteins.

Materials and Methods

Partially purified soybean hemagglutinin was prepared from untoasted soybean flour (Soyafluff 200W, Central Soya, Chicago, IL) according to the method of Liener (11). Final purification was achieved by chromatography on hydroxylapatite, prepared

[1] Current address: Research Division, Science Park, Smithville, TX 78957.

according to Tiselius et al. (12). The freeze-dried protein was
dissolved in Ca^{+2}- and Mg^{+2}- free phosphate buffered saline (CMF-
PBS), pH 7.5, and insoluble residue removed by centrifugation.
Protein concentrations were determined using an extinction coefficient of 1.28 at 280 nm for a 0.1% solution of SBA in a 1.0 cm
cuvette (5). Polyacrylamide slab gel electrophoresis of SBA was
performed in 10% acrylamide gels in the presence of 0.1% sodium
dodecyl sulfate (SDS). Samples in 2% SDS and 5% β-mercaptoethanol
were heated at 100° C for 2 min. prior to application to the gel.
Staining was carried out with 0.05% Coomassie Brilliant Blue R-250
in 7% acetic acid. The specific hemagglutination activity of SBA
was determined with rabbit erythrocytes by the method of Smith
et al. (13). One hemagglutination unit (HAU) is defined as the
minimum amount of lectin necessary to agglutinate erythrocytes.
The ability of saccharides to interact with the lectin was determined by their inhibition of lectin-induced agglutination of rabbit erythrocytes, according to the method of Smith et al. (13).
One hemagglutination inhibition unit (HAIU) is defined as the minimum amount of saccharide necessary to inhibit completely three
hemagglutination units of lectin. The purest available sugars
were obtained from various sources: 2-acetamido-2-deoxy-D-glucose
(GlcNAc), GalNAc, and 2-acetamido-2-deoxy-D-mannose (ManNAc) from
Sigma Chemical Co., St. Louis, Missouri, and D-Gal and lactose
from Fisher Scientific Co., Fair Lawn, New Jersey.

CD recordings were made on a Durrum-Jasco Model CD-SP Dichro-
graph, improved by D. P. Sproul of Sproul Scientific Instruments,
Tucson, Arizona. The sensitivity scale setting was 2 x 10^{-5}
dichroic absorbance per 1 cm on the recorder chart. Spectra were
measured at protein concentrations of 0.48 mg/ml (1.0-cm cell) in
the region above 250 nm and 0.046 mg/ml (0.1-cm cell) below
250 nm. A mean residue weight of 109 was calculated from the
amino acid analysis of SBA (5). These data are expressed in terms
of mean residue ellipticities [θ], in degrees·cm·dmol^{-1}. All
recordings were performed in CMF-PBS, pH 7.5, at 25 $\pm$ 2° C and
were repeated two or three times. CMF-PBS was prepared according
to Cronin et al. (14). A DuPont Model 310 curve resolver was used
to resolve CD curves into gaussian bands.

Results and Discussion

Preparation and Characterization of SBA. Partially purified
SBA was isolated from 700 g of untoasted soybean flour essentially
as described by Liener (11) following extraction of lipids with
petroleum ether. The procedure was carried out through the step
involving dialysis against 60% ethanol at -15° whereupon a precip-
itate formed. This precipitated crude hemagglutinin (2 grams) was
submitted to chromatography on hydroxylapatite as described by Lis
et al. (3). A summary of the purification of SBA is presented in
Table I. Lis et al. (15) reported the possible presence of isolectins of SBA based on its resolution into multiple components by

chromatography on DEAE-cellulose. Although it was possible to
obtain chromatographic patterns similar to those reported by Lis
and co-workers, homogeneity of these isolectins could not be
demonstrated by rechromatography under the same conditions.

TABLE I

PURIFICATION OF SBA

Purification Step	Weight Recovery mg	Specific Activity HAU/mg	Total Activity HAU x 10^{-6}
H_2O Extract	170,000	600	100
40-70% $(NH_4)_2SO_4$ ppt.	5,700	6,500	37
60% ethanol dialysis ppt.	2,000	12,000	25
Hydroxylapatite chromatography – active peak	290	80,000	23

SBA, prepared as described herein, possessed an ultraviolet
(UV) absorption spectrum and an extinction coefficient comparable
to those previously reported (5). Polyacrylamide gel electrophoresis in the presence of SDS resolved SBA into two closely spaced
peptide bands comparable to those reported by Lotan et al. (6).
SBA exhibited a specific activity of 80,000 HAU/mg.

Secondary Structure of SBA. SBA showed a negative CD band
centered at 225 nm and a positive band at 197 nm (Fig. 1a). Resolution of this curve into gaussian bands yielded maxima at 197,
217, 226, and 233 nm with [θ] values of 8900, -2400, -2900, and
-900, respectively. The amount of β structure (26%) was estimated
using the band at 217 nm, according to Chen et al. (16), taking
the value of -9200 for the β standard. No second Cotton effect or
trough could be discerned in the region of 208 nm indicating the
absence of any appreciable α-helical conformation. However, the
positive band at 196-198 nm is characteristic of a high content of
the pleated sheet conformation (17-19). Assessment of the
β-structural content of this protein must be taken with some
reservation due to the general uncertainty in quantitation of this
conformation from CD spectra. It is known that the optical activity of the β pleated sheet depends on the length and width of the
sheet (18) as well as solvent effects. Moreover, the CD effects
of peptide groups that are neither in helical or β pleated sheet
regions are little investigated. According to CD data the lectins

isolated from Canavalia ensiformis (7), Dolichos biflorus (8),
Pisum sativum (20), Robinia pseudoacacia (8), Ricinus communis
(9), and Bandeiraea simplicifolia (21) have a high content of the
β conformation. The lectin from Triticum vulgaris appears to be
an exception having only 12% β structure (10).

CD Band Fine Structure of SBA in the Near UV. Figure 1b
shows the CD spectrum of SBA in the 250-320 nm spectral zone.
This region is characterized by a small negative band at
300-310 nm, positive peaks at 294 and 288 nm, and a broad positive
region at 265-285 nm. There are crossover points at 300 and 259
nm. The bands at 300-310 and 294 nm are probably due to tryptophan
while the band at 288 nm is due to the tyrosine chromophore (22).
The broad region (265-285 nm) is characteristic of the overlapping
vibronic vicinal interactions of the aromatic chromophores (22).
This near UV spectrum is very similar to that of the α-D-galacto-
pyranosyl-binding lectin isolated from Bandeiraea simplicifolia
seeds (21). Since the soybean lectin is devoid of cystine (3,5,
15) all of the CD bands in the near UV region arise from vicinal
effects of the aromatic chromophores.

Saccharide Specificity of SBA. The saccharide specificity of
SBA was determined by hemagglutination inhibition assay (13). Of
the saccharides tested, GlcNAc and ManNAc exhibited less than 2
HAIU/μmol. However other saccharides tested possessed the following hemagglutination inhibitory activities: lactose, 20; D-Gal,
30; and GalNAc, 1000 HAIU/μmol. The relative activities are similar to those reported by Lis et al. (4) and suggest that an equatorial 2-acetamido group and an axial 4-OH group of the galacto-
pyranosyl ring are important in the binding of saccharides to the
lectin. Furthermore, the fact that disaccharides of GalNAc and
D-Gal were not significantly better inhibitors than the corresponding monosaccharides (4) suggests that the saccharide binding
region of SBA may be no larger than the size of a monosaccharide.

Effects of Saccharides on Conformation of SBA. None of the
saccharides investigated affected the CD spectrum of SBA in the
far UV (190-250 nm) spectral zone. However significant saccharide-induced effects were observed at 265-290 nm, the region in
which the aromatic chromophores display their Cotton effects.
These observations are in accord with the established idea that
weak aromatic transitions can gain or lose intensity through
vibronic vicinal interactions which practically do not affect the
backbone peptide chromophores. The results are compiled in Table
II.
     Conformational transitions were induced by GalNAc and to a
much lesser extent by lactose and D-Gal. Several concentrations
of GalNAc were utilized to establish the minimal saccharide concentration required to produce the maximal conformational transition. GalNAc at a concentration 1 mM induced maximal transitions

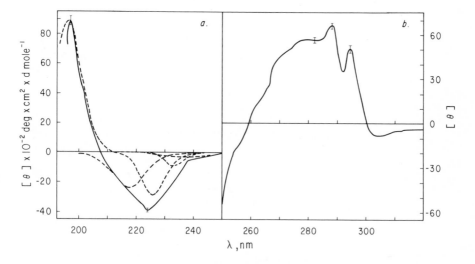

Figure 1. CD spectrum of SBA in the (a) far-UV and (b) near-UV regions. The lectin concentration was 0.046 mg/ml in the far-UV region and 0.48 mg/ml in the near-UV region. The light path length was 1 mm below 250 nm and 1 cm above 250 nm. The solid curves were constructed from four recordings each. The far-UV region was resolved into gaussian bands using the curve resolver. Solvent, CMF-PBS, pH 7.5. Bars indicate maximum deviation from mean. Comparable CD patterns were obtained using SBA, prepared by the method of Lotan et al. (6), obtained commercially from Miles Laboratories, Elkhart, Indiana.

in SBA as evidenced by alterations in the CD spectrum at 260-290
nm (Fig. 2). The saccharide effect was most pronounced at 282 nm
where SBA showed a mean residue ellipticity $[\theta]$ in degrees·cm^2·
dmol^{-1} of 56 $\pm$ 2 in the absence of saccharide. A $[\theta]$ value of
44 $\pm$ 2 was obtained in the presence of 1 mM GalNAc. These experiments were performed under conditions in which there were available 125 molecules of saccharide per saccharide binding site.
Addition of 0.1 mM GalNAc, 25 mM lactose, or 25 mM D-Gal to SBA
yielded $[\theta]$ values of 50, 49, and 50, respectively. See Table II.
The addition of 25 mM GlcNAc or ManNAc, which do not bind to SBA,
did not influence lectin conformation (Table II). By assuming
that the lectin binding sites are saturated at those saccharide
concentrations which yield the maximal effect on the CD spectrum,
it was possible to utilize the CD data to calculate an association
constant for the GalNAc/SBA interaction (7 x 10^3 liter·mole^{-1}).
This value compares reasonably well to that reported by Lotan
et al. (5) using equilibrium dialysis and gel filtration (3 x 10^4
liter·mole^{-1}).

TABLE II

SACCHARIDE-INDUCED ALTERATIONS IN CD OF SBA[a]

Saccharide	Conc.[b] (mM)	$[\theta]_{sugar}$ / $[\theta]_{no\ sugar}$			
		294 nm	288 nm	282 nm	252 nm
GlcNAc	25	1.00	1.02	1.00	0.97
ManNAc	25	1.00	1.06	1.04	1.00
Lactose	25	0.94	0.91	0.88	0.97
D-Gal	25	0.94	0.88	0.89	0.96
GalNAc	0.01	0.94	0.95	0.98	0.97
	0.1	0.96	0.92	0.89	0.97
	1	0.94	0.85	0.73	1.00
	25	0.92	0.86	0.76	1.00

a) The effects are expressed as ratios of $[\theta]_{sugar}/[\theta]_{no\ sugar}$
   at 294, 288, 282, and 252 nm. Maximum deviation from mean 4%.
   In all recordings the baselines were run with solutions of
   the saccharides in CMF-PBS, pH 7.5. The saccharides exhibited no CD bands in the near UV.
b) Concentration of the saccharide in the lectin-saccharide
   solution. In all cases, the concentration of the lectin
   was 0.48 mg/ml (4μM). Solvent, CMF-PBS, pH 7.5.

We conclude that the interaction of the lectin with GalNAc, D-Gal, or lactose resulted in an alteration of the asymmetric environment of aromatic side-chain chromophores present at the surface of the lectin molecule. According to the CD data (Table II and Fig. 2), the most likely chromophores which are involved in the SBA-saccharide interactions are tyrosine and tryptophan side chains. The CD changes at 282 and 288 nm probably involve tyrosine and those at 294 nm, tryptophan. Similar effects on aromatic side chain chromophores have been seen during saccharide binding to the lectins from Canavalia ensiformis (7) and Triticum vulgaris (10). Since there are overlaps of the various chromophore effects over the whole near-UV zone, a more definitive assignment of the CD bands requires additional investigation.

Acknowledgements

This work was supported by grants from The Robert A. Welch Foundation (G-051), The George and Mary Josephine Hamman Foundation and The Paul and Mary Haas Foundation.

Abstract

Conformational studies on soybean agglutinin (SBA) were performed using circular dichroism (CD). SBA exhibited a CD spectrum characterized by a small negative band at 300-310 nm, positive peaks at 294 and 288 nm, a broad positive region at 265-285 nm, a negative band centered at 224 nm and a positive peak at 197 nm. Analysis of the far ultraviolet CD bands indicated approximately 26% pleated sheet (β) and no evidence of α-helix. 2-Acetamido-2-deoxy-D-galactose (GalNAc) at a concentration 1 mM induced maximal conformational transitions in SBA (4µM) as evidenced by alterations in the CD spectrum at 260-290 nm. The saccharide effect was most pronounced at 282 nm where SBA showed a mean residue ellipticity $[\theta]$ in degrees·cm^2·dmol^{-1} of 56 ± 2 in the absence of saccharide. A $[\theta]$ value of 44 ± 2 was obtained in the presence of 1 mM GalNAc. Addition of 0.1 mM GalNAc, 25 mM lactose, or 25 mM D-galactose to SBA (4µM) yielded $[\theta]$ values of 50, 49, and 50, respectively. 2-Acetamido-2-deoxy-D-glucose and 2-acetamido-2-deoxy-D-mannose, which do not bind to SBA, did not influence lectin conformation. According to the CD spectra, the polypeptide chain backbone of the lectin was not affected by interaction with the saccharides.

Literature Cited

1. Sharon, N. and Lis, H.   Science (1972) 177, 949-959.
2. Liener, I. E. and Pollansch, M. J.   J. Biol. Chem. (1952) 197, 29-36.
3. Lis, H., Sharon, N., and Katchalski, E.   J. Biol. Chem. (1965) 241, 684-689.

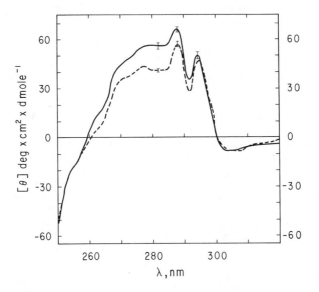

Figure 2. Effect of GalNAc on the conformation of SBA. The optical path length was 1 cm, the concentration of the lectin was 0.48 mg/ml, and the concentration of the saccharide was 1 mM. (———) lectin without sugar; (– – –) lectin with sugar. Solvent, CMF-PBS, pH 7.5. Bars indicate maximum deviation from mean.

4. Lis, H., Sela, B-A., Sachs, L., and Sharon, N. Biochim. Biophys. Acta (1970) 211, 582-585.
5. Lotan, R., Siegelman, H. W., Lis, H., and Sharon, N. J. Biol. Chem. (1974) 249, 1219-1224.
6. Lotan, R., Cacan, R., Cacan, M., Debray, H., Carter, W. G., and Sharon, N. FEBS Letters (1975) 57, 100-103.
7. Pflumm, M. N., Wang, J. L., and Edelman, G. M. J. Biol. Chem. (1971) 241, 4269-4370.
8. Père, M., Bourrillon, R., and Jirgensons, B. Biochem. Biophys. Acta (1975) 393, 31-36.
9. Shimazaki, K., Walborg, E. F., Jr., and Jirgensons, B. Arch. Biochem. Biophys. (1975) 169, 731-736.
10. Thomas, M. W., Walborg, E. F., Jr., and Jirgensons, B. Arch. Biochem. Biophys. (1977) 178, 625-630.
11. Liener, J. E. J. Nutr. (1953) 49, 527-539.
12. Tiselius, A., Hjerten, S., and Levin, O. Arch. Biochem. Biophys. (1956) 65, 132-155.
13. Smith, D. F., Neri, G., and Walborg, E. F., Jr. Biochemistry (1973) 12, 2111-2118.
14. Cronin, A. P., Biddle, F., and Saunders, F. K. Cytobios (1970) 2, 225-231.
15. Lis, H., Fridman, C., Sharon, N., and Katchalski, E. Arch. Biochem. Biophys. (1966) 117, 301-309.
16. Chen, Y-H., Yang, J. T., and Chau, K. H. Biochemistry (1974) 13, 3350-3359.
17. Jirgensons, B. "Optical Activity of Proteins and Other Macromolecules", (1973) 2nd ed., pp. 77-122, Springer-Verlag, Berlin.
18. Woody, R. W. Biopolymers (1966) 8, 669-683.
19. Balcerski, J. S., Pysh, E. S., Banora, H. M., and Toniolo, C. J. Am. Chem. Soc. (1976) 98, 3470-3473.
20. Bures, L., Entlicher, G., and Kocourek, J. Biochim. Biophys. Acta (1972) 285, 235-242.
21. Lönngren, J., Goldstein, I. J., and Zand, R. Biochemistry (1976) 15, 436-440.
22. Strickland, E. H. CRC Crit. Rev. Biochem. (1974) 2, 113-174.

RECEIVED September 8, 1978.

7

Dependence of Agglutination on Concanavalin A Molecular Transition

D. W. EVANS and P. Y. WANG

Institute of Biomedical Engineering, University of Toronto,
Toronto, Ontario M5S 1A4, Canada

Concanavalin A (Con A) has been reported to interact with
specific membrane saccharide receptors and agglutinate a wide
variety of transformed cells at low concentrations of the lectin
without affecting the normal untransformed parent line (1,2).
As a result of extensive studies, various theories have been proposed (2,3), however, the basis for the difference in agglutinability is not yet fully established.

The complexity and incomplete understanding of the cell surface make the interpretation of results from cell agglutination
studies very difficult at present. Another problem is in dis-
tinguishing between the effect of assay conditions on cell properties and on the Con A molecule. For example, it has been found
that the phase transition temperature for many membrane lipids is
around 15°C(4). Below this temperature there is a shift from a
fluid to a semicrystalline phase. Therefore, the loss of agglutinability of many transformed cells at low temperature has been
suggested as resulting from a decrease in mobility of lectin
receptors which may affect the tendency for them to cluster in
this semi-solid lipid matrix (2,5). However, others (6-8) have
proposed recently that this loss of agglutinability may be
explained by the transition of the Con A molecule to an inactive
form under these conditions. Since changes in cell surface properties undoubtedly occur, it is extremely difficult to distinguish between the two effects and determine their relative importance.

We have found that crosslinked dextran gel spheres (9) provide a useful model system with which to study agglutination.
One grade of these gel spheres can be extensively agglutinated by
Con A, whereas another grade, while still binding the lectin, is
not affected. In many ways the agglutination behaviour of this
system resembles that of transformed and normal mouse 3T3 cells
(10). Since unlike the cells, experimental conditions routinely
employed have minimal effect on the properties of the gel spheres,
changes in the agglutination behaviour may be related primarily
to effects on the Con A molecule.

0-8412-0466-7/79/47-088-076$05.00/0
© 1979 American Chemical Society

Materials and Methods

Con A, twice crystallized in saturated sodium chloride solution, was obtained from Miles Laboratories (Elkhart, Ind.). Fragmented sub-units were removed by gel filtration on Bio-Gel P-100 (11). Solutions were made up in Dulbecco's phosphate buffered saline (PBS) pH 7.4, and the concentration routinely determined by UV spectrophotemetric analysis based on $E_{280}^{1\%}$ (1 cm path) = 13.0.

Dextran Gel Spheres. Crosslinked dextran gel spheres (Sephadex), grade G-200, were obtained from Pharmacia (Montreal). A modified Neubauer hemocytometer was used at 40x magnification to obtain the number and size range of gel spheres in suspension. The depth of the counting chamber was increased from the usual 0.1 mm for routine cell counting to 0.3 mm to ensure unrestricted influx of the larger gel particles. A well-shaken 1:1 (v/v) suspension of pre-swelled gel spheres in PBS was diluted 5-fold and sampled with a 1-ml plastic serological pipet (Falcon, Oxnard, Ca.). The pipet had been pre-coated by drawing and discharging a gel sphere suspension several times, followed by rinsing 3 times with PBS. The number of spheres counted in the hemocytometer was subsequently converted to concentration (number of spheres per ml). The size distribution of the hydrated gel spheres was determined by measuring the diameter of each sphere in the sample with the aid of the hemocytometer grid squares of known dimension.

Agglutination Assay. A volume of 1 ml of PBS and 0.5 ml of Con A solution containing 0.5 to 4 mg/ml of lectin was added to 35 by 10 mm Petri dishes (Falcon Plastics, Oxnard, Ca.), and incubated at the assay temperature. A volume of 0.2 ml of a well-shaken, 1:1 (v/v) suspension of pre-swelled gel spheres in PBS at the required temperature was then added to each dish. The assay suspension was shaken horizontally on a Dubnoff metabolic shaker at 30 oscillations/min and an amplitude of 7 cm. Agglutination was scored visually at 40x magnification under the microscope: 0, no agglutination; +, slight; ++, low; +++, medium; ++++, extensive (12). Inhibition of agglutination was studied using the chaotropic agent sodium thiocyanate in the range 0-2 M. The pH of the assay solution was adjusted using 0.1 N HCl (pH 3), PBS (pH 7.4) and 0.05 M potassium carbonate-potassium borate-potassium hydroxide buffer (pH 10).

Con A Binding Assay. For each binding experiment a number of 60 by 15 mm Petri dishes containing 3.0 ml PBS and 1.5 ml of a 3 mg/ml Con A solution were adjusted to the required conditions of temperature or pH. A volume of 0.6 ml of a well-mixed 1:1 (v/v) suspension of dextran gel spheres was added with a 1 ml plastic pipet to each dish, followed by incubation with shaking. Control samples contained methyl α-$\underline{D}$- glucopyranoside (α-MG) at a final concentration of 0.1 M. At various intervals the contents of a sample and a control dish were centrifuged quickly and the supernatant withdrawn for UV spectrophotometric analysis at 280 nm

(1 cm path). Samples were taken until constant optical density
(O.D.) readings were obtained. The amount of Con A remaining in
solution was calculated from the final constant O.D. reading, and
the amount of Con A bound to the gel spheres was then obtained by
difference with the value for the control (13).

   Analysis of Con A by Ultracentrifuge. Sedimentation velocity
experiments were conducted at 60,000 rpm in a Beckman model E
analytical ultracentrifuge equipped with Schlieren optics. Con A
solutions were prepared at concentrations of 1, 5, 10, and 15 mg/
ml in PBS (pH 7.4) and run against pure solvent as base line in
double-sector 12 mm charcoal-filled epon cells. Temperatures
were controlled with the rotor temperature indicator control
(RTIC) unit. After attaining the desired speed, photographs of
the moving boundary were taken every 16 min. The Schlieren patterns were analysed on a Nikon model 6C shadowgraph comparator.
Bimodal patterns were resolved into two peaks and the peak mid-
points used for sedimentation velocity calculations. The relative areas under the peaks were determined by polar planimeter
measurement of the traced patterns. Sedimentation coefficients
were corrected for the effects of temperature and solvent (14)
and expressed relative to water at 20°C ($s_{20,w}$). A constant partial specific volume of 0.73 ml/g was used in these calculations
(15). The sedimentation coefficient of Con A (5 mg/ml) in 1 M
sodium thiocyanate was also determined at 20°C. After analysis,
the sample was dialysed against PBS to remove all thiocyanate and
the sedimentation coefficient redetermined.

Results

   The sampling of the gel sphere suspension using a pre-coated
plastic serological pipet was found to yield reproduceable results. The 1:1 (v/v) suspension, chosen to give an optimal number
of spheres for observation under agglutination assay conditions,
contained 4.95×10^5 spheres/ml. After correction for dilution,
a final concentration of 5.82×10^4 spheres/ml was obtained in the
assay solution. From the size distribution of the spheres determined in the hemocytometer, it was found that the diameters of
over 70% of the spheres lay in the range 100-220 µ. With an average sphere diameter of 160 µ, an average surface area of $8.04 \times
10^{-4}$ cm^2 was obtained.

   The agglutination of gel G-200 was dependent on Con A concentration (Figure 1). The maximum extent (++++) was observed at Con
A concentrations of 588 µg/ml and above. No further enhancement
was found with up to 10 mg Con A/ml. Agglutination was completely
inhibited by the presence of 0.1 M α-MG in the assay solution. When
gel G-200 spheres were incubated with 882 µg Con A/ml in the presence of increasing amounts of sodium thiocyanate, the agglutination decreased until complete inhibition was obtained at a final
concentration of 1 M NaSCN (Figure 2). No such change in agglutination was observed in the presence of sodium chloride. The

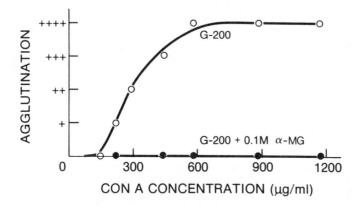

Figure 1.   *Agglutination of gel G-200 spheres by Con A at 20°C*

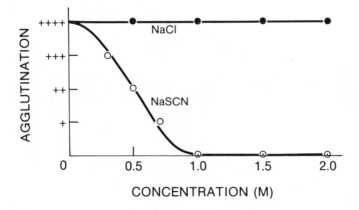

Figure 2.   *Effect of sodium thiocyanate and sodium chloride on Con A aggluti-nation of G-200 spheres*

results were identical whether the sodium thiocyanate was present
initially in the assay solution or if it was added to disperse
spheres already extensively (++++) agglutinated. The effect of
temperature on gel G-200 agglutination is shown in Figure 3. At
temperatures below 4^OC, no agglutination was observed with 882
μg Con A/ml. As the temperature was raised, the spheres started
to agglutinate, reaching a maximum (++++) above 15^OC. This temperature effect was reversible. Spheres agglutinated at 20^OC
(++++) were found to disperse on transfer to low temperature and
re-associate upon warming. As seen in Figure 4, the agglutination
was also pH dependent with complete inhibition at pH values less
than 5 or greater than 9. At about pH 6, there was a marked increase in agglutination reaching a maximum (++++) in the range
6.5-8 before an equally marked drop.

The binding of Con A to gel G-200 spheres under various conditions was determined in the presence of 882 μg/ml of the lectin.
This value was chosen because it was found to give a maximum
extent of agglutination (Figure 1). The binding was a gradual
process with about 2.7h required to obtain a constant level at
20^OC. In these experiments, 3.02 mg of an initial 4.5 mg of Con
A was bound at steady state (Table I). Based on a molecular
weight of 100,000 (16), this is equivalent to 6.13 x 10^{10} Con A
molecules bound per sphere or 7.62 x 10^{13} molecules per cm^2 of
the sphere surface. Other binding studies over a range of concentrations (Scatchard plots), have shown that at saturation the
actual number of Con A binding sites is about three times this
value (unpublished results). The presence of α-MG (0.1 M) in the
solution completely prevented binding. Further, Con A previously
bound to the steady state level could be completely and immediately released by the addition of 0.1 M α-MG. Since no change
in the Con A concentration in solution could be detected even
after 24 hr when α-MG was present, Con A does not appear to enter
the interior of the gel. Therefore, Con A must be bound primarily
on the surface of the sphere through interaction with the α-D-
glucopyranosyl residues of the dextran chains. In the presence of
1 M sodium thiocyanate the gel spheres still bound 2.94 mg Con A
at steady state (Table I). Similarly, low temperature had little
effect on the amount of Con A bound. As is shown in Figure 5, pH
has only a small effect on Con A binding in the range 4-8; however,
above this value there is a sharp drop with no Con A bound at pH
9 and over.

Ultracentrifugal analysis of Con A in solution revealed two
components, the relative proportions of which depended on the
experimental conditions. In Figure 6 (A), the Schlieren pattern
of Con A at a concentration of 5 mg/ml in PBS at 5^OC shows these
two components. An $s_{20,w}$ value of 5.52 S was obtained for the
fast component and 4.15 S for the slower component. By comparing
the relative areas under the peaks it was found that at this temperature 66% of the Con A molecules sedimented as the faster 5.52
S species. Higher temperatures favoured this species (Table II),

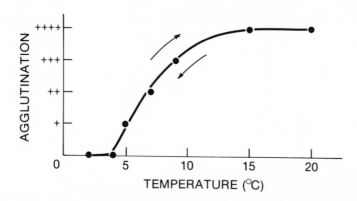

Figure 3.   Effect of temperature on Con A agglutination of G-200 spheres

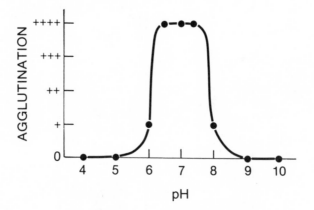

Figure 4.   Effect of pH on Con A agglutination of G-200 spheres

Table I

Con A bound to G-200 spheres under various conditions

Condition	O.D. at steady state (280nm)	change in O.D. units (280nm)	Con A bound (mg)	Con A molecules bound	
				per sphere ($\times 10^{-10}$)	per cm^2 sphere surface ($\times 10^{-13}$)
PBS	0.40	0.77	3.02	6.13	7.62
0.1 M α-MG	1.17	0	0	0	0
1 M NaSCN	0.42	0.75	2.94	10.80*	13.43*
4°C	0.43	0.74	2.90	7.54+	9.38+

* assuming a molecular weight of 55,000
+ assuming a molecular weight of 78,000

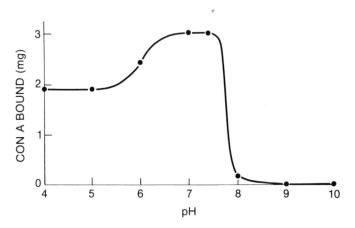

Figure 5. *Effect of pH on Con A binding of G-200 spheres*

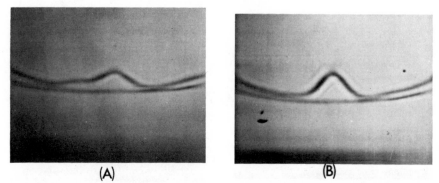

Figure 6. *Schlieren patterns of Con A (5 mg/ml) in PBS, pH 7.4: (A) 4°C; (B) 20°C. Movement is from left to right, speed 60,000 rpm, phase plate angle of 50°.*

until at 20°C essentially all of the molecules were in this form.
As seen in Figure 6 (B),

Table II

Effect of temperature on relative amounts of Con A
components in PBS at 5 mg/ml

Temperature (°C)	Percentage of 5.52 S species
5*	55
5	66
10	74
15	80
20	90

* concentration of 1 mg/ml

some of the 4.15 S component was still present at 20°C as indicated by the slight shoulder on the trailing edge of the peak.
The slower 4.15 S species was favoured by low concentration and at
1 mg/ml the relative amounts of the two components were about
equal (Table II). In 1 M sodium thiocyanate at 20°C, Con A
(5 mg/ml) sedimented as a single broad peak as seen in Figure
7 (A), with an $s_{20,w}$ value of 4.14 S. After dialysis, Figure 7
(B), the familar bimodal pattern returned with most of the Con A
present as the faster component. This peak gave an $s_{20,w}$ value
of 5.86 S.

Discussion

        Studies of Con A induced agglutination of live cells are sus-
ceptible to complications such as irreversibility of the agglutinated cells upon aging (17), endocytosis of bound lectin (18) and
effects of experimental conditions such as temperature, pH and
ionic strength on membrane properties (7). Alternatively, the
use of aldehyde-fixed cells (19), heatkilled cells (20), enucle-
ated cells (21) and glycoprotein-incorporated liposomes (22) may
lead to less problems being encountered.
        The crosslinking of bacterial formed strands of the poly-
saccharide dextran with epichlorohydrin, produces water-insoluble
gels which are available in the form of spheres. Various grades
of these gels differ only in the extent of the crosslinking reaction, and hence their swellability in water (9). The physical
and chemical properties of the gels are well documented (9,23)
and the binding of Con A to the gel spheres has been reported to
involve the α-D-glucopyranosyl units in the dextran chains (24).
The gel spheres have been used extensively in the isolation of
Con A by affinity chromatography (25), and a galactose-derivitized

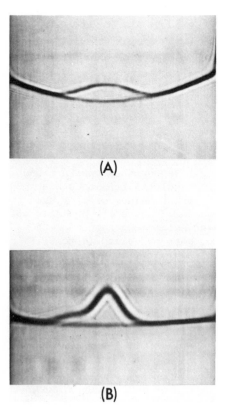

(A)

(B)

Figure 7. Schlieren patterns of Con A (5 mg/ml) in sodium thiocyanate at 20°C: (A) in 1 M NaSCN; (B) after dialysis against PBS. Movement is from left to right, speed 60,000 rpm, phase plate angle of 50°.

form has been shown to promote specific adhesion of transformed fibroblasts in vitro (26). However their potential use as a system to study agglutination has not been fully exploited (27,28).

In the presence of 882 μg Con A/ml, gel G-200 binds 7.62 x 10^{13} Con A molecules /cm^2 at steady state (Table I), and is extensively agglutinated (Figure 1). Since the polysaccharide chains are linked together by epichlorohydrin reaction, the gel contains no freely mobile Con A receptors. Therefore the agglutinability of these particles cannot be satisfactorily explained by lectin receptor clustering as has been proposed for cells (29). Rather, the agglutination behaviour may be related to the ability of the gel G-200 spheres to undergo transient deformation during collision in suspension, which may effect better contact and allow sufficient Con A molecules to abridge receptor sites on opposing surfaces thus holding the spheres together (30). Gel G-75 spheres have comparatively more rigid characteristics which may be responsible for their non-agglutinability. A Con A-glycogen complex used to compensate for insufficient contact, has produced extensive agglutination of gel G-75 spheres (31). The Con A-glycogen complex can also agglutinate untransformed and glutaraldehyde-fixed mouse 3T3 cells (10).

The agglutination behaviour of gel G-200 is temperature and pH dependent. At 20°C, maximum (++++) agglutination occurs at pH values between 6.5 and 8 (Figure 4). At pH values greater than 8, Con A does not bind to the gel spheres (Figure 5), probably due to formation of the high molecular weight aggregates of the lectin observed previously (15). At lower pH values (4 to 6), there is also a marked loss in agglutinability; however, in this case the binding is not appreciably affected. Con A has been shown to exist purely as a dimer in the pH range 3.5 to 5.8 (16,32). In this form the molecule retains its binding capacity without effecting agglutination (33). Above pH 5.8, Con A undergoes reversible association to form a tetramer (16). Therefore, the marked increase in agglutination of gel G-200 spheres above pH 6, corresponds to this transition of Con A from the inactive dimer form to the active tetrameric species.

At pH 7.4 in the presence of 882 μg Con A/ml, maximum agglutination of gel G-200 spheres is obtained at temperatures above 15°C (Figure 3). The decrease in the extent of agglutination at lower temperatures is not related to a change in the amount of Con A bound (2.90mg at 4°C compared to 3.02mg at 20°C – Table I). It is interesting to note that the change in agglutination behaviour occurs around 15°C which is similar to the finding in cell studies (5). Since these temperatures are not anticipated to have any adverse effects on the dextran gel, this change in agglutination of the G-200 spheres must again be related to transition of the Con A molecule.

The two components revealed by ultracentrifugal analysis of Con A in solution (Figure 6) correspond to the dimer and tetramer forms of the molecule. The $s_{20,w}$ values of 4.15 S and 5.52 S respectively, obtained in this study, are consistent with previous reports (15,16). The relative proportions of the two components were found to be temperature dependent. At 20°C, when gel G-200 spheres are extensively agglutinated, essentially all the Con A molecules are in the tetramer form (Table II). The amount of dimer increases with decreasing temperature. When the extent of agglutination is greatly reduced as at about 5°C (Figure 3), approximately equal amounts of the two components (55% tetramer) are present for a Con A concentration of 1 mg/ml. These results have also been observed by Gordon and Marquardt (6). They have reported that at a concentration of 1.2 mg/ml and a pH of 7.2, Con A is approximately 95% tetramer at 22°C, and about 60% at 4°C. Another report (7) using a much lower concentration (0.05-0.2 mg/ml) has indicated that at pH 7.2 and 4°C only 10% tetramer is present. However, McCubbin and Kay (34) have shown that the molecular weight of Con A is concentration dependent, and at pH 7 low concentration may favour the dimer. This effect was also observed in the present study (Table II).

Therefore, similar to the effect of pH discussed previously, the loss of agglutinability of gel G-200 spheres at low temperature, follows the conversion of Con A molecules from the active tetramer to the inactive dimer. The reversible nature of this transition is demonstrated by the observation that spheres agglutinated at 20°C quickly disperse on transfer to low temperatures and re-associate when the temperature condition is restored (Figure 3).

The importance of temperature in the stabilization of the Con A tetramer strongly indicates the involvement of hydrophobic interactions in this process (16,35). Similarly, in the case of antigen-antibody reactions, hydrophobic interaction has been suggested to play an important role in the stabilization of multivalent antibodies (36). Hydrophobic interaction is affected by chaotropic ions such as thiocyanate (37), and in the presence of 1 M sodium thiocyanate there is complete inhibition of agglutination of the gel G-200 spheres (Figure 2). In this situation the chaotropic agent may have two possible effects. It may interfere with the stabilization of the Con A tetramer or it may affect the interaction of Con A with any hydrophobic sites on the dextran gel spheres. The latter possibility follows from the observation that a number of low molecular weight aromatic substances are retarded during gel filtration on Sephadex columns (38). This effect is reported to be most pronounced with the heavily crosslinked gels G-25, G-15 and in particular G-10, and is much less significant for gel G-200. However, our results show that in the presence of 1 M sodium thiocyanate, Con A is still bound to the spheres (Table I) which contradicts a hydrophobic involvement in the Con A-dextran gel interaction, especially since the bound Con A can be completely and immediately released by the addition of α-MG. Therefore, it appears that the chaotropic agent must have acted by disrupting hydrophobic interactions stabilizing the

Con A tetramer. This has been confirmed by ultracentrifugal analysis. In the presence of 1 M sodium thiocyanate a single component is observed (Figure 7) with an $s_{20,w}$ value of 4.14 S. The
existence of Con A in the dimer form under these conditions
accounts for the binding without inducing agglutination. After
dialysis to remove all thiocyanate, the original bimodal Schlieren pattern returns with essentially all the Con A molecules once
again in the tetramer form. The dialysed Con A agglutinates gel
G-200 spheres to the same extent as before (++++), which indicates
that the only effect of the chaotropic agent is in the reversible
transition of the tetramer to dimer with subsequent change in
agglutination.

The use of dextran gel spheres allows the effects of experimental conditions on agglutination to be studied relatively free
of complications due to undefined changes in system properties.
Variations in the ability of Con A to induce agglutination under
different conditions as aforementioned, are attributed to the
tetramer-dimer transition of the Con A molecule. Lack of agglutination is directly correlated with the conversion of Con A from
the active tetramer form to the inactive dimer species. The
results of this study support earlier contentions (6-8) that the
effects of temperature on cell agglutination cannot be interpreted
exclusively in terms of changes in cell surface properties.

Acknowledgement

This work was supported by a grant from the Ontario Cancer Treatment and Research Foundation.

Literature Cited

1. Inbar, M. and Sachs, L., Proc. Natl. Acad. Sci. U.S. (1969),
   ·63, 1418-1425.
2. Rapin, A.M.C. and Burger, M.M., Adv. Cancer Res. (1974), 20
   1-91.
3. Nicolson, G.L., Intl. Rev. Cytol. (1974), 39, 89-190.
4. Reinert, J.C. and Steim, J.M., Science (1970), 168, 1580-
   1582.
5. Noonan, K.D. and Burger, M.M., J. Cell Biol. (1973), 59,
   134-142.
6. Gordon, J.A. and Marquardt, M.D., Biochim. Biophys. Acta
   (1974), 332, 136-146.
7. Huet, M., Eur. J. Biochem. (1975), 59, 627-632.
8. Huet, C., Lonchamp, M., Huet, M. and Bernadac, A., Biochim.
   Biophys. Acta (1974), 365, 28-39.
9. Sephadex Gel Filtration in Theory and Practice pp. 4-9,
   Pharmacia Canada, Dorval, Quebec, (1974).
10. Wang, P.Y. and Evans, D.W., to be published.
11. McKenzie, G.H. and Sawyer, W.H., J. Biol. Chem. (1973), 248,
    549-556.

12. Sela, B., Lis, H., Sharon, N. and Sachs, L., J. Membrane Biol. (1970), 3, 267-279.
13. Uchida, T. and Matsumoto, T., Biochim. Biophys. Acta (1972), 257, 230-234.
14. Van Holde, K.E., in "The Proteins", (H. Neurath and R.L. Hill, eds.), 237, Academic Press, New York, (1965).
15. Agrawal, B.B.L. and Goldstein, I.J., Arch. Biochem. Biophys. (1968), 124, 218-229.
16. McKenzie, G.H., Sawyer, W.H. and Nichol, L.W., Biochim. Biophys. Acta (1972), 263, 283-293.
17. Nicolson, G.L., Ser. Haemat. (1973), 3, 275-291.
18. Noonan, K.D. and Burger, M.M., J. Biol. Chem. (1973), 248, 4286-4292.
19. Marquardt, M.D. and Gordon, J.A., Exp. Cell Res. (1975), 91 310-316.
20. De Petris, S., Raff, M.C. and Mallucci, L., Nature New Biology (1973), 244,275-278.
21. Wise, G.E. and Larsen, R., Exp. Cell Res. (1976), 97, 141-150.
22. Juliano, R.L. and Stamp, D., Nature (1976), 261, 235-237.
23. Gelotte, B. and Porath, J., in "Chromatography", (E. Heftman, ed.), 2nd. ed., 343, Reinhold, New York, (1967).
24. Agrawal, B.B.L. and Goldstein, I.J., Biochem. J. (1965), 96, 23C.
25. Agrawal, B.B.L. and Goldstein, I.J., Biochim. Biophys. Acta (1967), 147, 262-271.
26. Chipowski, S., Lee, Y.C. and Roseman, S., Proc. Natl. Acad. Sci. U.S. (1973), 70, 2309-2312.
27. Inoue, M., Mori, M., Utsumi, K. and Seno, S., Gann (1972), 63, 795-799.
28. Rutishauser, U. and Sachs, L., Proc. Natl. Acad. Sci. U.S. (1974), 71, 2456-2460.
29. Nicolson, G.L. Nature (1972), 239, 193-197.
30. Evans, D.W., Ph.D. Thesis, University of Toronto, (1977).
31. Wang, P.Y. and Evans, D.W., Federation Proceedings (1977), 36, 795.
32. Kalb, A.J. and Lustig, A., Biochim. Biophys. Acta (1968), 168, 366-367.
33. Gunther, G.R., Wang, J.L., Yahara, I., Cunningham, B.A. and Edelman, G.M., Proc. Natl. Acad. Sci. U.S. (1973), 70, 1012-1016.
34. McCubbin, W.D. and Kay, C.M., Biochem. Biophys. Res. Comm. (1971), 44, 101-109.
35. Kauzman, W., Advan. Protein Chem. (1959), 14, 1-20.
36. Kleinschmidt, W.J. and Boyer, P.D., J. Immunol. (1952), 69, 247-255.
37. Dandliker, W.B. and DeSaussure, V.A., in "The Chemistry of Biosurfaces", (M.L. Hair, ed.), Vol. 1, pp. 1-43, Marcel Dekker Inc., New York, (1972).
38. Determann, H. and Walter, I., Nature (1968),219, 604-605.

RECEIVED September 8, 1978.

8

Antibodies Against Oligosaccharides

DAVID A. ZOPF, CHAO-MING TSAI, and VICTOR GINSBURG

National Institute of Arthritis, Metabolism, and Digestive Diseases,
National Institutes of Health, Bethesda, MD 20014

The carbohydrate chains in membrane glycoproteins and glycolipids occur in small amounts and are difficult to isolate and
characterize chemically. As an alternative to direct chemical
analysis we have been developing immunologic methods to detect
defined carbohydrate sequences on cell surfaces by taking advantage of the fact that many of the sugar sequences that occur in
the complex carbohydrates of cell surfaces also are present in
the free oligosaccharides of human milk (1). When coupled to
carrier polypeptides, these oligosaccharides are immunogenic and
antibodies that specifically bind defined carbohydrate structures
can be obtained (2,3).

For example, the human Le^b blood group-active hapten lacto-N-
difucohexaose I (see Table I for structures of oligosaccharides)
was coupled to poly-L-lysine by a modification of the method of
Arakatsu, et al. (4). Le^b blood group activity of the synthetic
glycoconjugate which contained 61 sugar haptens per mole of poly-
L-lysine (approx. MW = 70,000) was demonstrated by quantitative
precipitin reaction with standard anti-Le^b serum (kindly provided
by Dr. Donald Marcus) and specific inhibition of the precipitin
reaction with the free Le^b blood group hapten (Figure 1). When
the same glycoconjugate was complexed with succinylated hemo-
cyanin and administered to a goat in complete Freunds adjuvant,
anti-Le^b antibodies were obtained (2). Hapten binding specificity
of the antiserum can be determined using a radioimmunoassay described in Figure 2. Of the oligosaccharide haptens tested, the
Le^b-active hapten lacto-N-difucohexaose I is the most active,
giving 50% inhibition of binding at .25 μM. Its monofucosyl
analogues lacto-N-fucopentaose I and lacto-N-fucopentaose II are
approximately 100-fold less active and the tetrasaccharide,
lacto-N-tetraose, is inactive at 2.5 mM. These results indicate
that the fucose residues of lacto-N-difucohexaose I are necessary
to provide the best fit with the anti-Le^b immunoglobulin combining site. The requirement for the fucose residues could be explained in two ways: 1) their presence tends to favor some
critical conformation necessary for binding of the tetrasaccharide

TABLE I

lacto-<u>N</u>-tetraose (LNT)	$\overset{\beta}{}$ $\overset{\beta}{}$ $\overset{\beta}{}$ Gal 1→3 GlcNAc 1→3 Gal 1→4 Glc
lacto-<u>N</u>-fucopentaose I (LNF I)	$\overset{\beta}{}$ $\overset{\beta}{}$ $\overset{\beta}{}$ Gal 1→3 GlcNAc 1→3 Gal 1→4 Glc 2 ↑ α Fuc 1
lacto-<u>N</u>-fucopentaose II (LNF II)	$\overset{\beta}{}$ $\overset{\beta}{}$ $\overset{\beta}{}$ Gal 1→3 GlcNAc 1→3 Gal 1→4 Glc 4 ↑ α Fuc 1
lacto-<u>N</u>-difucohexaose I (LND I)	$\overset{\beta}{}$ $\overset{\beta}{}$ $\overset{\beta}{}$ Gal 1→3 GlcNAc 1→3 Gal 1→4 Glc 2 4 ↑ α ↑ α Fuc 1 Fuc 1
lacto-<u>N</u>-neotetraose (LNnT)	$\overset{\beta}{}$ $\overset{\beta}{}$ $\overset{\beta}{}$ Gal 1→4 GlcNAc 1→3 Gal 1→4 Glc
lacto-<u>N</u>-fucopentaose III (LNF III)	$\overset{\beta}{}$ $\overset{\beta}{}$ $\overset{\beta}{}$ Gal 1→4 GlcNAc 1→3 Gal 1→4 Glc 3 ↑ α Fuc 1

Abbreviations used in the figures are given in parentheses.
All sugars except fucose (Fuc) are of the <u>D</u>-configuration.

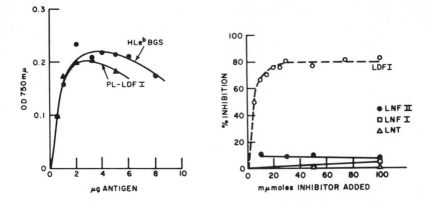

Figure 1.   (Left) Precipitation of goat Le^b antibody by the polylysine-lacto-N-difucohexaose I conjugate (PL-LDF I) and by soluble human Le^b blood group substance (HLe^bBGS). The reaction mixture contained 50 μl of goat serum and the indicated amount of antigen in a total volume of 250 μl. (right) Hapten inhibition of the precipitin reaction between the polylysine-lacto-N-difucohexaose I conjugate and goat Le^b antibody. The reaction mixture contained 50 μl of goat serum, 3 μg of the conjugate, and the indicated amount of inhibitor in a total volume of 250 μl. The abbreviations used are given in Table I.

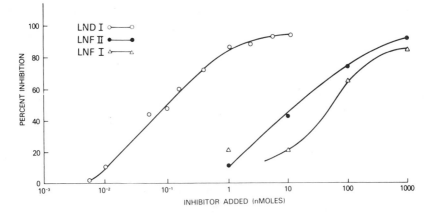

Figure 2.   Inhibition of antibody binding of ³H-lacto-N-difucohexaitol I by oligosaccharides. Immune goat serum, 10 μl, is mixed with varying amounts of inhibitor in Tris buffer, pH 7.5, containing 0.14 M NaCl, 5 × 10⁻⁴ M MgSO₄, and 1.5 × 10⁻⁴ CaCl₂ in a final volume of 300 μl. After incubation for 20 min at 37°, 6.7 pmoles of ³H-lacto-N-difucohexaitol I (10⁵ cpm) in 100 μl of buffer are added. After a second incubation at 37° for 30 min the mixture is passed through a nitrocellulose filter. The filter is washed with 10 ml of the same buffer and counted in a scintillation counter. In the absence of inhibitor 2200 cpm are bound. When the assay is performed with preimmune serum the filter nonspecifically traps 30 to 40 cpm. The abbreviations used are given in Table I.

chain of lacto-N̲-tetraose to the antibody combining site; or
2) the fucosyl residues themselves interact directly with the
immunoglobulin combining site. Evidence to be presented below
tends to favor the latter possibility.

The polylysine-oligosaccharide conjugates used in the
experiments just described elicit useful antibodies, but they
produce high-titer antisera in only a few animals. The relatively low immunogenicity of these conjugates prompted a search
for a procedure that would make more efficient use of scarce
oligosaccharides by employing a high yield coupling reaction to
form the glycoconjugates as well as a more immunogenic carrier
to increase the fraction of animals responding. Using the
methods that follow we have produced and characterized antibodies
against several milk oligosaccharides starting with less than
100 μmoles of each oligosaccharide. Useful antibody responses
were obtained in more than 80 percent of rabbits immunized.

In the first step of the coupling procedure the reducing
sugar residue reacts with the alkylamino group of β-(p̲-amino
phenyl)-ethylamine to form a Schiff base (Figure 3). Reduction
with sodium borohydride gives an N̲-alkyl-1-amino-1-deoxyalditol
derivative (5̲). The derivative is obtained in 50 to 95 percent
yields based on starting sugar and the reaction proceeds under
mild conditions so that acid-labile sugar linkages are not degraded. These derivatives are easily diazotized to the hemp
seed protein edestin, previously shown by Himmelspach and
Kleinhammer to be an efficient immunogenic carrier for sugar-
flavazole derivatives (6̲). Conjugates prepared by this method
contain 30 to 40 sugar haptens per molecule of edestin (MW =
310,000). A typical immune response of a rabbit immunized with
an edestin conjugate containing lacto-N̲-tetraose is shown in
Figure 4.

Specificity of antibody binding can be determined by comparing inhibitory activities of structurally related oligosaccharides in radioimmunoassay. From preliminary calculations
we estimated from the number of counts bound by the antiserum
on day 74 (Figure 4) that the K_a for binding of ^{3}H-lacto-N̲-
tetraitol is at least 10^6 liter/mole. However, inhibition by
lacto-N̲-tetraose gives a value about 100-fold lower, that is,
the apparent K_a for lacto-N̲-tetraose is 2 x 10^4 (Figure 5). The
reduction to lacto-N̲-tetraitol increases inhibitory activity
about 500-fold (apparent K_a = 1.6 x 10^7). As the phenethylamine
derivative of lacto-N̲-tetraose is only 2-fold more active than
lacto-N̲-tetraitol, it appears that the non-carbohydrate portion
of the derivative contributes very little to antibody binding.
Inhibition by the various unreduced oligosaccharides is difficult
to interpret as they inhibit only at high concentrations.

Clearly, the antibodies formed against the oligosaccharide
hapten in this case recognize the modified glucose linkage arm
present in the sugar derivative used to prepare the immunogen.
Why the antibodies produced are specific for the reduced

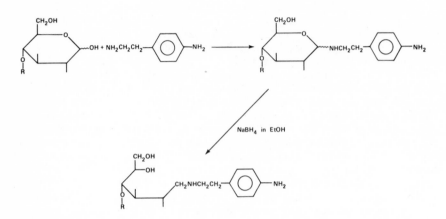

Figure 3.   Reaction of sugars with β-(p-aminophenyl)-ethylamine

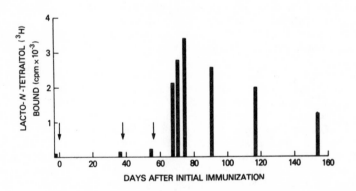

Figure 4.   Binding of ³H-lacto-N-tetraitol by rabbit serum (R31) in response to immunization with edestin-φetNH-lacto-N-tetraose. Antigen in complete Freunds adjuvant was administered on Days 1, 38, and 58 (arrows). The binding assay is performed as described in the legend to Figure 2.

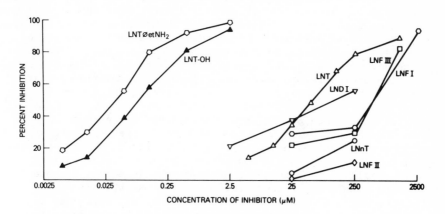

Figure 5.   *Inhibition of anti-lacto-N-tetraose (R31) by oligosaccharides. Inhibition assays are performed as described in the legend to Figure 2. Abbreviations of oligosaccharides are given in Table I. An abbreviation followed by -OH or ϕetNH₂ refers to the reduced form or the β-(p-aminophenyl) ethylamine derivative of the oligosaccharide, respectively.*

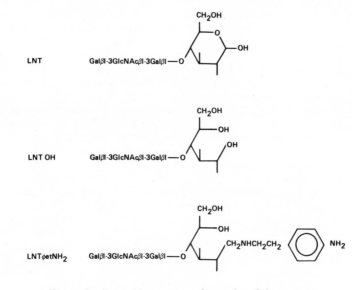

Figure 6.   *Lacto-N-tetraose and its reduced derivatives*

oligosaccharide can be understood by comparing the structures of
native lacto-N-tetraose with lacto-N-tetraitol and its
phenethylamine derivative   (Figure 6).   Antibodies were formed
against an immunogen containing the phenethylamine derivative of
lacto-N-tetraose diazotized to protein.   Apparently the reduced
form of glucose present in the immunogen constitutes an important
immune determinant in the rabbit.   Lacto-N-tetraitol has the same
structure and therefore binds equally well.   In contrast, most
lacto-N-tetraose molecules in solution have the ring form of
glucose.

     The importance of the non-reducing terminal structure
Galβ1-3GlcNAc can be seen by comparing activities of lacto-N-
tetraose with lacto-N-neotetraose which is identical except that
galactose is linked to the 4 position of N-acetylglucosamine.
The possibility that lacto-N-neotetraose binds poorly because of
steric hindrance by a group at the 4 position is rendered unlikely by the fact that lacto-N-fucopentaose II, which has fucose
on the 4 position of N-acetylglucosamine, inhibits.   Substitution
on the 2 position of galactose also results in only an 8-fold
loss in binding activity indicating that the major interactions
between antibody and sugar are not disturbed.   In fact, significant binding occurs when two fucose residues are present as
in lacto-N-difucohexaose I.   Therefore, the low binding activity
of lacto-N-neotetraose is not due to steric hindrance by
galactose at the 4 position of N-acetylglucosamine but rather
the absence of a critical substituent in the 3 position.   Sur-
prisingly it appears that fucose can partially substitute for
galactose in that binding activity actually increases when
lacto-N-neotetraose is substituted with fucose in the 3 position
of N-acetylglucosamine to form lacto-N-fucopentaose III.   Some
molecular models of these oligosaccharides which account for
these findings are shown in Figure 8.

     In the upper left is lacto-N-tetraose.   Lacto-N-fucopentaose
I and lacto-N-fucopentaose II contain the identical tetraose
structure with single fucose residues added in the 2 position of
galactose or the 4 position of N-acetylglucosamine, respectively.
The sugar sequence, GlcNAcβ1-3Galβ1-4Glc, common to all oligosaccharides is oriented in the same way in each picture and a
dashed line has been drawn to define a plane normal to the page.
Approach of antibody molecules to the atoms above this plane
would not be hindered by either the 2- or 4-substituted fucosyl
residues, both of which project below the plane.   Loss of the 3-
linked terminal galactose in lacto-N-neotetraose leaves an empty
space above the plane as the 4-linked galactose projects below
the plane.   These models account for the binding data accumulated for these oligosaccharides as well as other data regarding
their chromatographic properties and their differential rates of
cleavage by glycosidases (7,8).

     The pattern of binding of various complex sugars with anti-
lacto-N-tetraose and anti-lacto-N-difucohexaose I raises an

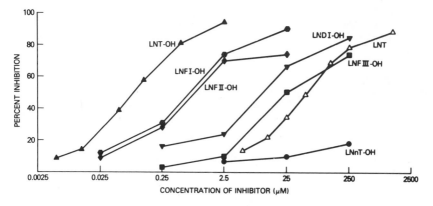

Figure 7.   *Inhibition of anti-lacto-N-tetraose (R31) by oligosaccharides. Conditions for the inhibition assay are given in the legend to Figure 2. Abbreviations are given in Table I and in the legend to Figure 5.*

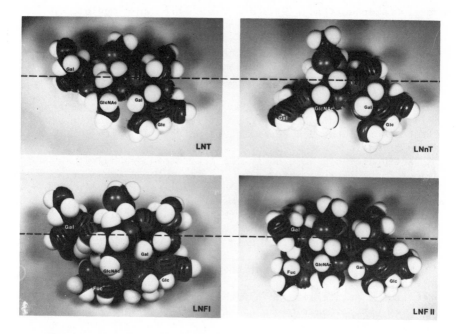

Figure 8.   *Molecular models of oligosaccharides. The line indicates the position of an arbitrary plane normal to the page above which lacto-N-tetraose (LNT) is unaltered by substitution of fucosyl residues to form lacto-N-fucopentaose I (LNF I) or lacto-N-fucopentaose II (LNF II). In lacto-N-neotetraose (LNnT) nonreducing terminal galactose projects below this plane   (see Table I for the structure of the oligosaccharides).*

interesting point: the anti-lacto-N̲-difucohexaose I antibody
recognizes lacto-N̲-tetraose carrying two fucosyl residues.
Activity decreases significantly when either one of the fucosyl
residues is absent and activity is completely lost when both
fucoses are missing. On the other hand, the rabbit anti-
lacto-N̲-tetraose recognizes the tetrasaccharide alone and its
binding activity decreases only slightly when one or both fucosyl
residues are added to the chain. Thus both antisera can be
inhibited by the Le[b] hapten, lacto-N̲-difucohexaose I. The best
interpretation of this data is that the anti-lacto-N̲-difuco-
hexaose I antibody recognizes a surface of the lacto-N̲-difuco-
hexaose I molecule that includes the fucose residues, whereas
the anti-lacto-N̲-tetraose antibody recognizes the opposite
surface which is not significantly altered by substitution with
fucose as illustrated diagramatically in Figure 9.

The possibility that antibodies can bind the same oligosaccharide but from different sides might explain the puzzling
fact that antibodies with the same chemical specificity, as judged
by hapten binding or hapten inhibition studies, sometimes have
different serologic specificities. For example, a rabbit anti-
paragloboside antibody (9) and a Waldenström cold agglutinin
(cold agglutinin McC) (10) are both inhibited by paragloboside,
yet have different serologic specificities: The rabbit antibody
reacts equally well with human cord and adult erythrocytes (9)
while the cold agglutinin reacts strongly with cord cells but
weakly or not at all with adult cells (10). If the erythrocyte
receptors were actually substituted paraglobosides and the two
proteins bound to different sides of the paragloboside sugar
chain, the antibodies would react differentially with the substituted paraglobosides depending on which side of the chains
the substitutions occurred.

Antibodies can be used for detection of carbohydrate sequences in biological materials. For example, a mannotetraose
(Manα1-3Manα1-2Manα1-2Man) obtained by selective acetolysis of
yeast mannan was derivatized with phenethylamine and diazotized
to edestin (3). Specificity of the anti-mannotetraose antibodies obtained was studied by the radioimmunoassay methods
just described. Results of these studies are summarized in
Table II.

The best inhibitor of antibody binding is mannotetraosyl-
phenethylamine. Mannotetraitol is 4-fold less active, again
indicating that the non-carbohydrate portion of the phenethylamine derivative contributes relatively little to immune binding. Compared to mannotetraitol with a K_a of 3.4 x 10^6 liter/
mole the mannotetraose has a K_a only about 10-fold lower. When
reducing terminal mannose is absent, the activity of mannotetraitol falls only about 2-fold, which indicates that antibody
binding is mainly directed against the non-reducing trisaccharide
terminal sequence. This idea gains further support from results
with the remaining inhibitors: modification of the mannotetraose

TABLE II

Inhibition of Anti-mannotetraose by Sugars

Sugar Inhibitor	Concentration Required for 50% Inhib. (μM)	Inhibition at 1 mM (%)
Man $1\overset{\alpha}{\rightarrow}3$ Man $1\overset{\alpha}{\rightarrow}2$ Man $1\overset{\alpha}{\rightarrow}2$ ManϕetNH$_2$	0.06	-
Man $1\overset{\alpha}{\rightarrow}3$ Man $1\overset{\alpha}{\rightarrow}2$ Man $1\overset{\alpha}{\rightarrow}2$ Mannitol	0.25	-
Man $1\overset{\alpha}{\rightarrow}3$ Man $1\overset{\alpha}{\rightarrow}2$ Man $1\overset{\alpha}{\rightarrow}2$ Man	2.8	100
Man $1\overset{\alpha}{\rightarrow}3$ Man $1\overset{\alpha}{\rightarrow}2$ Man	5	-
Man $1\overset{\alpha}{\rightarrow}3$ Man $1\overset{\alpha}{\rightarrow}2$ Man $1\overset{\alpha}{\rightarrow}2$ Man 3 $\uparrow\alpha$ Man 1	100	-
Man $1\overset{\alpha}{\rightarrow}2$ Man $1\overset{\alpha}{\rightarrow}2$ Man $1\overset{\alpha}{\rightarrow}2$ Man	1000	-
Man $1\overset{\alpha}{\rightarrow}2$ Man $1\overset{\alpha}{\rightarrow}2$ Man	> 1000	30
Man $1\overset{\alpha}{\rightarrow}2$ Man	> 1000	4
Man $1\overset{\alpha}{\rightarrow}3$ Man $1\overset{\beta}{\rightarrow}4$ GlcNAc	> 1000	23
Man $1\overset{\alpha}{\rightarrow}3$ Man $1\overset{\beta}{\rightarrow}4$ GlcNAc 2 $\uparrow\alpha$ Man 1	> 1000	0
Man $1\overset{\alpha}{\rightarrow}6$ Man $1\overset{\alpha}{\rightarrow}6$ Man	> 1000	12
Mannose	> 1000	0

Immune rabbit serum, 5 μl is mixed with varying amounts of inhibitor in buffer in a final volume of 300 μl. After incubation for 2 h at 25°, 6.7 pmole of [^{3}H]mannotetraose (10^3 cpm) in 100 μl of buffer is added and incubation is continued for 2 h at 25°. Binding of tritiated sugar is determined as described in Figure 2.

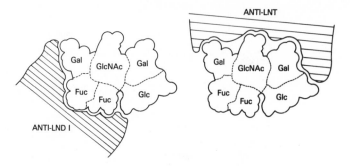

Figure 9. *Hypothetical scheme for binding of antibody against lacto-N-difuco-hexaose I (Anti-LND I) and antibody against lacto-N-tetraose (Anti-LNT) to opposite sides of the oligosaccharide lacto-N-difucohexaose I. Profiles of mono-saccharide units were drawn from a molecular model similar to those shown in Figure 8.*

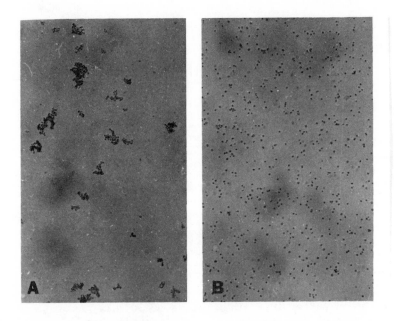

Figure 10. *Photomicrographs of a suspension of* Saccharomyces cerevisiae *incubated for 3 hr at 35° with rabbit antimannotetraose serum diluted 1:100 in buffered saline in the absence (A) or presence (B) of 1 mM mannotetraose*

chain by elongation with an additional non-reducing mannosyl residue reduces binding 20-fold and changing the non-reducing terminal linkage from Manα1-3 to Manα1-2 reduces binding by a factor of 200. Deletion of non-reducing terminal residues, alteration of the penultimate linkages, and changes in linkage positions result in almost complete loss of binding activity. As the antimannotetraose antibodies in this serum show specificity for the non-reducing trisaccharide sequence Manα1-3Manα1-2Man they can be used for detection of this sequence in biological material. For example, this antiserum specifically agglutinates whole yeast cells (Figure 10). Agglutination is inhibited as shown on the right by free mannotetraose but not by free mannose nor by other sugars we have tested shown in Table II.

In summary, it is possible to immunize animals with purified oligosaccharides coupled to proteins and obtain specific antisera that can be characterized by radioimmunoassay and which may prove useful for studying complex **carbohydrates.**

LITERATURE CITED

1) Kobata, A., in Methods in Enzymology (Ginsburg, V., ed.), Vol. 28, pp. 262-271, Academic Press, New York (1972).
2) Zopf, D.A., Ginsburg, A., and Ginsburg, V., J. Immunol. (1975) 115 1525-1529.
3) Zopf, D.A., Tsai, C., and Ginsburg, V., Arch. Biochem. Biophys. (1978) 185 61-71.
4) Arakatsu, Y., Ashwell, G., and Kabat, E.A., J. Immunol. (1966) 97 858-866.
5) Jeffrey, A.M., Zopf, D.A., and Ginsburg, V., Biochem. Biophys. Res. Commun. (1975) 62 608-613.
6) Himmelspach, K., and Kleinhammer, G., in Methods in Enzymology (Ginsburg, V., ed.), Vol. 28, pp. 222-231, Academic Press, New York, (1972).
7) Wiederschain, G.Y., and Rosenfeld, E.L., Biochem. Biophys. Res. Commun. (1971) 44 1008-1014.
8) Alhadeff, J.J., Miller, A.L., Wenaas, H., Vedrick, T., and O'Brien, J.S., J. Biol. Chem. (1975) 250 7106-7113.
9) Schwarting, G.A., and Marcus, D.M., J. Immunol. (1977) 118 1415-1419.
10) Tsai, C.-M., Zopf, D.A., Wistar, R., and Ginsburg, V., J. Immunol. (1976) 117 717-721.

RECEIVED September 8, 1978.

9

Interaction of Glycosyl Immunogens with Immunocyte Receptor Sites in the Synthesis of Anti-glycosyl Isoantibodies

JOHN H. PAZUR and KEVIN L. DREHER

Department of Biochemistry and Biophysics, The Pennsylvania State University, University Park, PA 16802

DAVID R. BUNDLE

Division of Biological Sciences, National Research Council, Ottawa, Canada, KIA OR6

Anti-glycosyl antibodies are defined as those antibodies which are induced by carbohydrate containing immunogens and which combine with a specific carbohydrate moiety of these immunogens (1). In chemical structure, the immunogens may be glycans (2-5), glycoproteins (6-9), glycolipids (10,11) or synthetic carbohydrate-protein conjugates (12-15). The immunogens of the glycan type are important components of the cell walls of pathogenic bacteria and such compounds form the basis of the scheme for the serological classification of these organisms (16). The immunogens of the glycoprotein type are important constituents of mammalian cells and are of great significance in the transformation of such cells into neoplasmic cells, particularly in the transformations brought about by viruses (17,18). The immune system is the primary means of defense against infections by bacterial pathogens and against the proliferation of neoplasmic cells of many types of carcinomas. The immunogens on the surface of these cells stimulate the immune system to produce antibodies which initiate the process that ultimately leads to the destruction of the bacterium or to the retardation of growth and in some cases the elimination of the neoplasmic cells.

Basic information on the biochemical reactions of the immune process is highly desirable in order that more effective methods of utilizing the immune system for the diagnosis and treatment of diseases may be developed. Many steps of the process are already reasonably well delineated, but the sequence of reactions and the biological mechanisms for some remain obscure (19). For example, it is well known that only certain structural features of the immunogen, the immunodeterminant groups, are involved in the stimulation of immunocytes to produce antibodies. These immunodeterminant groups are also the structural moieties of the antigens which combine with antibodies to form the antibody-antigen complex. However, it is not known how these groups interact with the receptor substances on the immunocytes or how this interaction initiates the sequence of biochemical reactions resulting in the synthesis of antibodies.

In antigens of the glycan and glycoprotein types the immunodeterminant groups are very often terminal carbohydrate moieties.
In such cases the antibodies which are elaborated against these
antigens are anti-glycosyl antibodies. Studies on the types of
anti-glycosyl antibodies induced by different antigens with the
same immunodeterminant groups should yield valuable information on
the immune process.

During the past few years we have been investigating the
nature of the anti-glycosyl antibodies induced in rabbits by
immunogens with many types of carbohydrate immunodeterminant
groups. In these studies rabbits have been immunized with
vaccines of bacterial glycosyl antigens or with synthetic carbohydrate-protein conjugates. Antisera have been obtained from
these rabbits after suitable periods of immunization by standard
methods of immunology. The types of anti-carbohydrate antibodies
in the sera have been deduced from data of hapten inhibition
experiments utilizing a variety of carbohydrates. Affinity
chromatography techniques have then been used to prepare anti-
glycosyl antibodies which are specific for a single type of
carbohydrate unit. Some of the chemical and immunological properties of such antibodies have been determined.

To date, antibodies of anti-galactose (20), anti-glucose
(21), anti-N-acetyl-glucosamine (22), anti-glucuronic acid (23),
anti-lactose (20) and anti-isomaltose (24) types have been
obtained. To obtain these antibodies different types of bacterial
antigens and carbohydrate-protein conjugates have been used, but
only three will be discussed in this report. One of these antigens
is a diheteroglycan of glucose and galactose in the cell wall of
Streptococcus faecalis and the other two are synthetic carbohydrate-protein conjugates, galactosyl bovine serum albumin (Gal-
BSA) and lactosyl bovine serum albumin (Lac-BSA). The methods for
the preparation of these antigens are described later. It should
be pointed out that all of these antigens possess the same types
of terminal carbohydrate moieties, galactosyl or lactosyl units,
which as will be seen, are the immunodeterminant groups.

The diheteroglycan of glucose and galactose is a type specific
carbohydrate in the cell wall of Streptococcus faecalis (5).
This glycan was extracted by a trichloroacetic acid method and
purified to homogeneity by alcohol precipitation and bio-gel
filtration methods. The complete molecular structure of the
glycan has been deduced from data of methylation analysis in
combination with enzymic and chemical degradation reactions (25,
26). A diagrammatic representation of the total structure for
this glycan is shown in Figure 1. The molecular weight of the
glycan is 15,000 and a typical molecule is composed of a main
chain of eighteen glucose-glucose-galactose repeating units and
eighteen disaccharide side chains attached to the central glucose
residue of each repeating unit. The disaccharide side chains are
mainly lactose units but a few are cellobiose units. As indicated
above the types of glycosidic linkages have been determined by

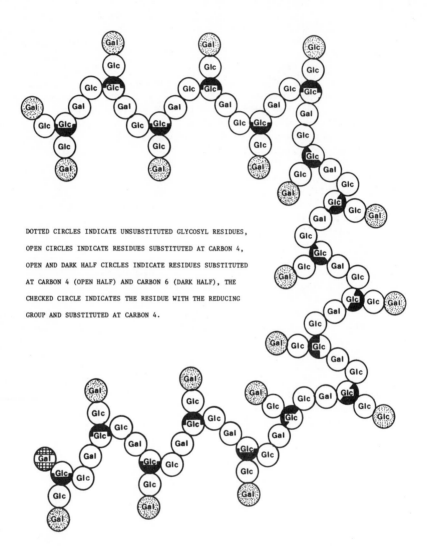

DOTTED CIRCLES INDICATE UNSUBSTITUTED GLYCOSYL RESIDUES,

OPEN CIRCLES INDICATE RESIDUES SUBSTITUTED AT CARBON 4,

OPEN AND DARK HALF CIRCLES INDICATE RESIDUES SUBSTITUTED

AT CARBON 4 (OPEN HALF) AND CARBON 6 (DARK HALF), THE

CHECKED CIRCLE INDICATES THE RESIDUE WITH THE REDUCING

GROUP AND SUBSTITUTED AT CARBON 4.

Carbohydrate Research

Figure 1. Diagrammatic representation of the structure of a typical molecule of the diheteroglycan of glucose and galactose from Streptococcus faecalis *(26)*

methylation analysis and the configuration of the linkages has been shown to be beta by enzymic tests. The experimental details of the structural studies have been published (25,26).

The carbohydrate-protein conjugates that have been employed in these studies are galactosyl bovine serum albumin (Gal-BSA) and lactosyl bovine serum albumin (Lac-BSA). The reaction sequence for the synthesis of these conjugates involves the preparation of the per-acetyl derivatives of galactosyl bromide or lactosyl bromide, reaction of either derivative with 8-ethoxycarbonyl octanol, deacetylation of the product, formation of the hydrazide, conversion to the azide and finally coupling the azide to the bovine serum albumin. Reaction conditions for the various steps of the synthesis are described in the literature (14,27) and the sequence is shown diagrammatically in Figure 2. Analysis of the conjugates for carbohydrate content by the phenol-sulfuric acid method (28) showed that about one third of the lysine residues of bovine serum albumin were bonded to carbohydrate residues.

Two types of vaccines and immunization regimes were employed for immunizing rabbits with these immunogens. The bacterial vaccine was a suspension of non-viable cells of Streptococcus faecalis with the glycan in situ in the cell wall. The details of the preparation of the vaccine have been described (25). Immunization was intraveneously three times per week for periods of up to fifteen weeks. Blood samples were collected weekly and antisera were prepared from these samples. Vaccines of Gal-BSA and Lac-BSA were prepared by mixing 0.2% solution of the conjugates in saline-phosphate buffer of pH 7.2 with an equal volume of complete Freunds adjuant. Immunizations with the latter vaccines were performed subcutaneously in the back of the neck three times in weekly intervals (27). Blood samples were collected 30 days after the beginning of immunization and antisera were prepared from these samples in the usual manner.

Agar diffusion patterns for the reactions of these antisera with the immunogens are shown in Figure 3, left plate. It will be noted that a strong precipitin band was obtained with each set of antigens and antisera. Hapten inhibition experiments with galactose, lactose and a variety of other carbohydrates showed that galactose and lactose were potent inhibitors of the precipitin reactions and thus all three antisera contained antibodies which combined with galactose or lactose units. These antibodies have been isolated by adsorption on lactosyl-sepharose and elution of the adsorbed antibodies with galactose and lactose solutions. Typical elution patterns for the three antisera samples and for a pre-immune serum from one of the rabbits are shown in Figure 4.

It will be noted in pattern A of Figure 4, that only serum protein was present in the pre-immune serum and antibodies were not eluted by galactose or lactose. Pattern B of the Figure shows that, in addition to the serum proteins, two 280 nm absorbing components were obtained from the anti-S. faecalis serum. One component eluted with galactose and the other eluted with lactose.

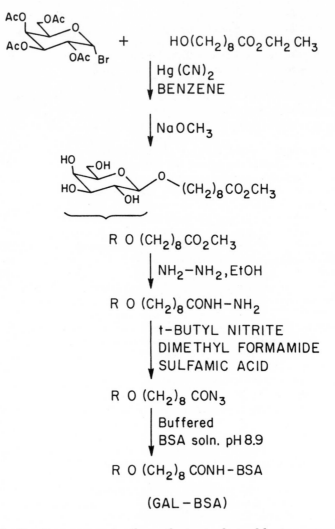

Figure 2.   *Reaction sequence for the synthesis of galactosyl bovine serum albu-min (Gal-BSA)*

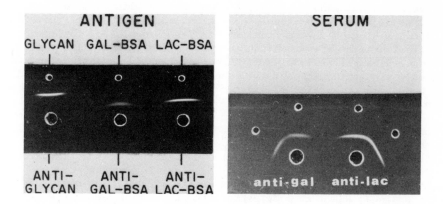

Journal of Biochemistry

Figure 3. *Agar diffusion plates: left plate as labeled, right plate contains anti-gal or anti-lac antibodies in the center wells and the diheteroglycan in the peripheral wells (29)*

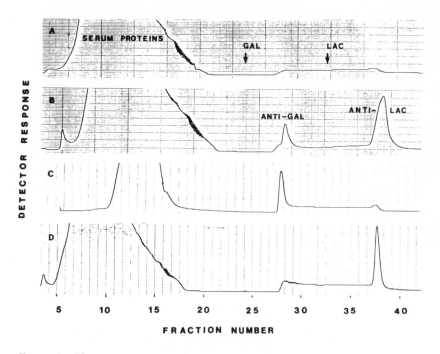

Figure 4. *Elution patterns for rabbit antisera samples from a lactosyl sepharose column: A = preimmune serum, B = anti-S -faecalis serum, C = anti-Gal-BSA serum and D = anti-Lac-BSA serum*

That both components were antibodies was established by agar
diffusion tests with these preparations and solutions of the
glycan performed in the usual manner. The results of the agar
diffusion tests are shown in Figure 3, right plate. The two
preparations of antibodies have been designated as anti-galactose
(anti-gal) antibodies and anti-lactose (anti-lac) antibodies and
some of the properties of the two sets of antibodies are described
in a later section.

The affinity chromatographic patterns from the antisera from
animals immunized with Gal-BSA and Lac-BSA are shown in patterns C
and D respectively of Figure 4. It will be noted that only one
set of antibodies is present in the antisera of animals immunized
with these immunogens. Thus only anti-gal antibodies were
obtained from the anti-Gal BSA serum and anti-lac antibodies from
the anti-Lac BSA serum. It should be pointed out again that all
of the immunogens possess common terminal carbohydrate units, β-D-
galactosyl units and two possess common terminal β-lactosyl units.
Therefore it was very surprising to find that the three immunogens
elicited different types of immune responses.

A number of properties of the anti-gal and anti-lac antibodies
obtained from the anti-S. faecalis serum have been determined.
Both of the preparations react in agar diffusion tests with the
glycan (Figure 3, right plate) and are therefore antibodies. Data
on hapten inhibition with various carbohydrates and the two antibody types are shown in Figure 5. It will be noted in the right
panel that the anti-lac antibodies are inhibited only by lactose
and lactose derivatives and not by galactose or compounds with
terminal galactose units. However the anti-gal antibodies are
inhibited by all compounds with terminal galactose units as seen
by the data in the left panel of Figure 5. Further, it should be
noted that a forty-fold difference in the concentration of the
inhibitors exists with the anti-lac antibodies being inhibited by
the lower concentration of inhibitors. These results establish
that one set of antibodies combines with galactose and the other
with lactose. The combination occurs even when the galactose or
lactose are terminal structural units of other compounds.

The sedimentation constants and molecular weights calculated
from ultracentrifugation data for both types of antibodies were 7s
and 150,000 respectively. Agar diffusion tests with goat antisera
against rabbit IgA, IgG and IgM showed that both antibody sets are
of the IgG immunoglobulin type (29).

When the anti-gal and the anti-lac antibodies were subjected
to gel electrophoresis an unexpected result was obtained. Such
gels showed that each preparation consisted of multi-protein
components with the anti-gal antibodies consisting of 6 proteins
and the anti-lac antibodies consisting of 12 different proteins.
Photographs of the gels for the two preparations are shown in
Figure 6. Gels stained with glycoprotein stains (30,31) showed
that all the components are glycoproteins.

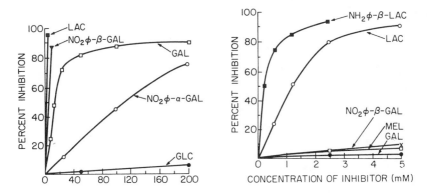

Journal of Biological Chemistry

Figure 5.   Quantitative inhibition curves for the precipitin reactions between the diheteroglycan and the anti-gal antibodies (left panel) or the anti-lac antibodies (right panel) (29)

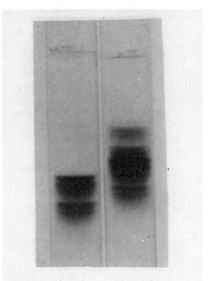

anti-gal   anti-lac

Biochemical and Biophysical
Research Communications

Figure 6.   Gel electrophoretic patterns for the preparations of anti-gal and anti-lac antibodies (1)

Since the components of each set were eluted by the same hapten group, galactose or lactose, and each protein combines with the same structural unit of the antigen, the individual proteins of each set have been designated as isoantibodies. This definition is a more restrictive definition of an isoantibody than is employed by immunologists (32) but is in line with the terminology employed by enzymologists for multi-molecular forms of enzymes (33).

The isoantibodies of the anti-gal and the anti-lac types have been separated into individual components by electrofocusing techniques. Results with the anti-gal isoantibodies are shown in Figure 7, top pattern. It will be noted that six distinct 280 nm absorbing peaks were obtained by the electrofocusing method. Gel electrophoresis patterns of the peak fractions for the anti-gal antibodies are also shown in Figure 7, bottom pattern. It will be noted that each component migrates as a single narrow band on electrophoresis. Since the original preparation of anti-gal isoantibodies on ultracentrifugation yield a symmetrical pattern (20), the isoantibodies are of the same molecular size. On the basis of these findings it is reasonable to conclude that the individual components are homogeneous antibodies.

The three antisera and the anti-glycosyl antibodies isolated from these antisera have been tested in agar diffusion tests for cross-reactivity with the three immunogens and with bovine serum albumin. These results are shown in Figure 8.

It will be noted in the top panels of Figure 8 that the original anti-Gal-BSA serum and the purified anti-gal antibodies from this serum reacted only with Gal-BSA but not with the other immunogens tested, Lac-BSA, glycan or BSA. As seen in the middle panels of the Figure, the anti-Lac-BSA serum reacted with Lac-BSA, Gal-BSA and BSA but the anti-lac antibodies from such antiserum reacted only with the Lac-BSA. Evidently this serum contained a population of antibodies which reacted with the BSA portion of the immunogens, thereby accounting for the reaction of this serum with Gal-BSA and BSA.

The anti-S. faecalis antiserum reacted with Lac-BSA and the glycan as shown in the bottom panels of Figure 8. The anti-lac antibodies from this serum also reacted with these two immunogens yielding a similar pattern as the unfractionated antiserum. It should be noted that the antiserum and antibody preparations exhibited partial identity with these two antigens. The anti-gal antibodies also reacted with Lac-BSA and the glycan but the precipitin bands were quite different.

The antisera from six rabbits immunized with vaccine of di-heteroglycan in situ in the cell wall, from three rabbits immunized with the Lac-BSA, and from two rabbits immunized with Gal-BSA vaccine have been analyzed by the above method. The results with all rabbits collaborate the findings presented in the foregoing. These findings can be interpreted as a manifestation of the mode of interaction of the immunodeterminant groups of the immunogens

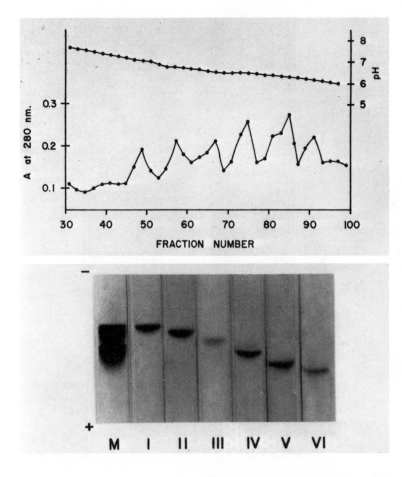

Biochemical and Biophysical Research Communications

Figure 7. *Electrofocusing pattern (top panel) and gel electrophoretic patterns (bottom panel) for the anti-gal isoantibodies (20)*

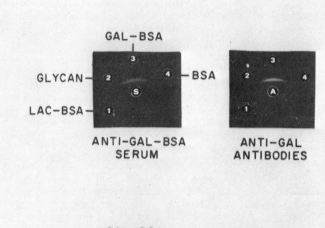

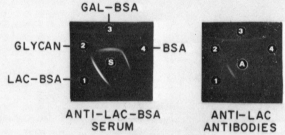

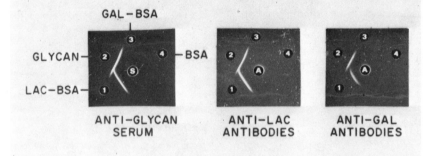

Figure 8.   Agar diffusion plates of various antisera and antibody preparations with the different immunogens

with the substances that compose the receptor sites of the immunocytes of the host. Two types of interactions will be considered
and the consequences of these interactions will be indicated.

First, in line with the current notion that a single immunocyte synthesizes a single type of protein (19), it is likely that
six different immunocytes are stimulated by the terminal galactose
units and twelve different immunocytes are stimulated by terminal
lactose units. Since each of the immunocytes is programmed to
synthesize a different and unique protein, six anti-gal proteins
and twelve anti-lac proteins will be produced. If this interpretation is correct then the isoantibodies of each set could be
significantly different in the basic structure of the polypeptide
chains. Studies are in progress on the determination of the
structural features of the isoantibodies with the view of obtaining evidence for or against this interpretation.

Second, the interaction of the immunogens and immunocytes may
be one in which the immunodeterminant group of the immunogen
combines with a single immunocyte. This single type of immunocyte
will multiply to produce a uniform population of the new immunocytes. These immunocytes produce a protein with the same basic
polypeptide structure but in the processing of the protein,
different amounts of carbohydrate residues or amide groups are
introduced into the protein. As a result the antibodies directed
against a single immunodeterminant group occur in multiple molecular forms, the isoantibodies.

If the first interpretation is the correct one, then in the
immunization with S. faecalis cells with the diheteroglycan of
glucose and galactose, one group of immunocytes recognizes terminal galactose units while another group recognizes terminal
lactose units. The result is the multiplication of these two
groups of cells and the synthesis of sets of anti-gal and anti-lac
isoantibodies. If the second interpretation is the correct one,
then only one immunocyte recognizes the galactose units and another
immunocyte recognizes the lactose units. In this case, one population of a single cell type leads to the synthesis of the set of
anti-gal isoantibodies and another population of a single cell
type leads to the synthesis of the set of anti-lac isoantibodies.

In the immunizations with the synthetic carbohydrate-protein
conjugates the galactosyl unit of Gal-BSA is responsible for the
stimulation of immunocytes leading to the production of anti-gal
isoantibodies, while with the Lac-BSA only the lactosyl moiety is
recognized by the immunocytes and only anti-lac isoantibodies are
produced. With the latter antigen anti-gal isoantibodies are not
produced even though the immunogen possesses terminal galactose
units. The differences in the responses to the three immunogens
with the same immunodeterminant groups can be interpreted to
indicate a role for cell surface topography and macromolecular
conformation of the immunogen in directing the synthesis of antibodies. Several recent reports (34-36) have appeared pointing to
possible roles for these structural features in the interaction of

immunodeterminant groups with the receptor substances on different
cell surfaces.

Literature Cited

1.   Pazur, J. H., Miller, K. B., Dreher, K. L. and Forsberg,
     L. S., Biochem. Biophys. Res. Commun., (1976), 70,
     545-550.
2.   Avery, O. T., Heidelberger, M., and Goebel, W. F., J. Exp.
     Med., (1925), 42, 709-725.
3.   Kabat, E. A., and Berg. D., J. Immunol., (1953), 70, 514-532.
4.   Krause, R. M. and McCarty, M., J Exp. Med., (1962) 115,
     131-140.
5.   Pazur, J. H., Anderson, J. S. and Karakawa, W. W., J. Biol.
     Chem., (1971), 246, 1793-1798.
6.   Gold, P., and Freedman, S. O., J. Exp. Med., (1965), 121,
     439-461.
7.   Terry, W. D., Henkart, P. A., Coligan, J. E. and Todd,
     C. W., Transplant. Rev., (1974), 20, 100-129.
8.   Hammarström, S., Engvall, E., Johansson, B. G., Svensson, S.,
     Sunblad, G. and Goldstein, I. J., Proc. Nat. Acad. Sci., USA,
     (1975), 72, 1528-1532.
9.   Kabat, E. A., in H. S. Isbell (Editor), "Carbohydrates in
     Solution, Advances in Carbohydrate Series," 334-361, American
     Chemical Society, Washington, D.C. (1973).
10. Siddiqui, B. and Hakomori, S., J. Biol. Chem., (1971), 246,
     5766-5769.
11. Sung, S. S. J., Esselman, W. J. and Sweeley, C. C., J. Biol.
     Chem., (1973), 248, 6528-6533.
12. Karush, F., J. Am. Chem. Soc., (1957), 79, 3380-3384.
13. Allen, P. Z., Goldstein, I. J., and Iyer, R. N., Biochemistry,
     (1967), 6, 3029-3036.
14. Lemieux, R. U., Bundle, D. R. and Baker, D. A., J. Am. Chem.
     Soc., (1975), 97, 4076-4083.
15. Zopf, D. A., Tsai, C. and Ginsburg, V., Arch. Biochem.
     Biophys., (1978), 185, In Press.
16. Lancefield, R. C., Harvey Lectures Ser., (1940-1941), 36,
     251-290.
17. Duff, R. and Rapp, F., J. Virol., (1971), 8, 469-477.
18. Hill, M., and Hillova, J., Adv. Cancer Res., (1976), 23,
     237-297.
19. Eisen, H. N. in B. D. Davis, R. Dulbecco, H. N. Eisen, H.
     S. Ginsberg and W. A. Wood (Editors), "Microbiology,"
     Chapters 14 and 17, Harper and Row Publishers, Hagerstown,
     MD (1973).
20. Pazur, J. H., and Dreher, K. L., Biochem. Biophys. Res.
     Commun., (1977), 74, 818-824.
21. Pazur, J. H., Cepure, A., Kane, J. A. and Karakawa, W. W.,
     Biochem. Biophys. Res. Commun., (1971), 43, 1421-1428.

22. Karakawa, W. W., Wagner, J. E. and Pazur, J. H., J. Immunol., (1971), 107, 554-562.
23. Pazur, J. H., Dropkin, D. J., Dreher, K. L., Forsberg, L. S., and Lowman, C. S., Arch. Biochem. Biophys., (1976), 176, 257-266.
24. Kane, J. A., Karakawa, W. W. and Pazur, J. H., J. Immunol., (1972), 108, 1218-1226.
25. Pazur, J. H., Cepure, A., Kane, J. and Hellerquist, C. G., J. Biol. Chem., (1973), 248, 279-284.
26. Pazur, J. H. and Forsberg, L. S., Carbohyd. Res., (1978), 58, In Press.
27. Lemieux, R. U., Baker, D. A., and Bundle, D. R., Can. J. Biochem., (1977), 55, 507-512.
28. Dubois, M., Gilles, K. A., Hamilton, J. D., Rebers, D. A. and Smith, F., Anal. Chem. (1956), 28, 350-356.
29. Pazur, J. H., Dreher, K. L., and Forsberg, L. S., J. Biol. Chem., (1978), 253, In Press.
30. Fairbanks, G., Steck, T. L. and Wallach, D. F. H., Biochemistry, (1971), 10, 2606-2617.
31. Eckhardt, A. E., Hayes, C. E. and Goldstein, I. J., Anal. Biochem., (1976), 73, 192-197.
32. Potter, M., Lieberman, R., and Dray, S., J. Mol. Biol., (1966), 334-346.
33. Commission on Biochemical Nomenclature, "Enzyme Nomenclature," pg. 32, Elsevier Scientific Publishing Co., Amsterdam, The Netherlands (1973).
34. Ramasamy, R., Immunochemistry, (1976), 13, 705-708.
35. Ostrand-Rosenberg, S., Immunogenetics, (1976), 3, 53-64.
36. Gurd, J. W., Biochemistry, (1977), 16, 369-374.

RECEIVED September 8, 1978.

10

The Specificity of Sugar Taste Responses in the Gerbil

WILLIAM JAKINOVICH, JR.

The Department of Biological Sciences, Herbert Lehman College,
The City University of New York, Bronx, NY 10468

Since sucrose is one of the most common soluble
sugars associated with other nutrients in plants and
seeds (1,2), the ability to survive may be linked to
an animal's ability to taste sucrose (3,4) - hence,
the evolution of a sucrose receptor site. Such an
idea has support because sucrose is the sweetest sugar
to humans (5,6), and by far the most uniformly prefer-
red by mammals (7,8). Moreover, it evokes in mammals
a greater gustatory nerve discharge than any other
sugar (3,9,10,11). This is a report of a detailed
physiological study of the sugar taste response in the
gerbil (12,13,14). For the gerbil, the best taste
stimuli are sugars which most closely resemble
sucrose.

Method

The Mongolian gerbil was chosen as the experimental animal because among mammals it possesses the
largest taste nerve response and lowest taste threshold to sucrose (3). The sugar taste response used was
the integrated (average) response of the taste neurons
since this is an index of the response of the entire
taste receptor population (15). The electrical activity was obtained by touching the chorda tympani nerve
of an anesthetized gerbil with a nichrome electrode
(100 μm diameter) connected to a differential amplifier.
The electrical activity, displayed on an oscilloscope,
could be monitored by a loudspeaker (Figure 1).

Stimulus Presentation. Distilled water was
allowed to flow continuously at a rate of 0.13-0.17
ml/sec over a gerbil's tongue, extended with a fine
fishhook. Test solutions (2-4 ml) were alternated
with the distilled water rinse without interruption of

0-8412-0466-7/79/47-088-116$05.00/0
© 1979 American Chemical Society

fluid flow. The temperature of the distilled water
rinse and of the taste solution were identical (25° ±
1°C). Sugar solutions were presented in a series of
steps increasing by approximately 1/2 log molar concentration, i.e., 0.0001 M, 0.0003 M, 0.001 M...0.1 M,
0.3 M. Each animal received a sucrose series and a
test compound series at least twice. The mean
responses were calculated and used in further analysis.
A standard solution (0.3 M sucrose) was presented frequently between test solutions. Whenever the standard
solutions elicited responses that varied more than
± 10%, all interjacent responses were rejected.

Sugars. Sugars were obtained from Pfanstiehl
Laboratories, Waukegan, Illinois and Sigma Chemical
Company, St. Louis, Missouri or were synthesized (12,
13,14).

Results

    When any effective chemical flowed onto the
tongue, there was an initial rapid rise of neural
activity which was dependent upon concentration. (A
response in this study was defined as the difference
between the integrated potential of the spontaneous
activity evoked by the water flow and the greatest
integrated potential elicited by a given solution
applied to the tongue.) Figure 2 is a typical series
of recordings. On semilogarithmic coordinates, the
concentration-response function for sugars was always
sigmoidal, similar to the sucrose concentration
response curve (Figure 3).

    Disaccharides. Of all the disaccharides tested,
sucrose was the best stimulus (Table I); it gave the
greatest response at all concentrations. All other
disaccharides tested had response thresholds 10 or
30 times higher than sucrose.

    Methyl Glycosides of D-Fructose and D-Glucose.
Reducing sugars mutarotate in aqueous solutions producing isomers which have different taste intensities
(16,17,18,19,20) and taste qualities (21,22); such
isomers tend to obscure taste-structure relationships.
Therefore, where possible, methyl glycosides of
D-fructose and D-glucose were compared. The two methyl
glycosides which resemble the moieties of sucrose most
closely, methyl α-D-glucopyranoside and methyl β-D-
fructofuranoside were the most stimulatory (Figure 4).

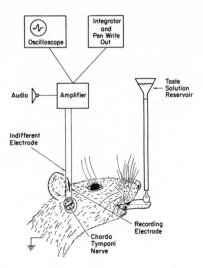

Figure 1.　*Schematic drawing of the experimental set up*

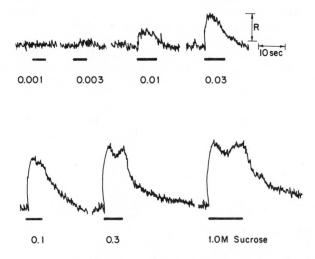

Brain Research

Figure 2.　*Integrated neural discharge from the gerbil's chorda tympani nerve in response to a series of increasing concentrations of sucrose applied to the tongue. The solid bars under the records indicate stimulus duration, R is the measure of response (12).*

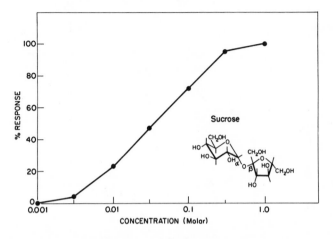

Figure 3.   Plot of neural discharge from Figure 2

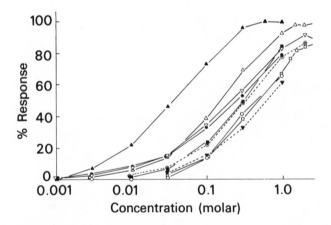

Figure 4.   Mean integrated response of the chorda tympani nerve discharge in the gerbil to sucrose (▲), methyl α-D-glucopyranoside (△), methyl β-D-glucopyranoside (□), methyl β-D-fructofuranoside (●), methyl α-D-fructofuranoside (○), methyl α-D-fructopyranoside (◼), methyl β-D-fructopyranoside (▼), 1,5-anhydro-D-mannitol (⊙), fructose (equilibrium mixture, 25°C) (▽). Responses are relative to the maximum sucrose response.

TABLE I

STIMULATING EFFECTIVENESS OF SOME DISACCHARIDES (MEAN VALUES)

Sugar	Structure*	Threshold** (molar)	CR_{50} (molar)	K_d (molar)	Maximum response	N
Sucrose	Glu α(1→2) Fru	0.001	0.042 ± 0.005^a	0.037	1.0	32
Turanose	Glu α(1→3) Fru	0.03	0.23 ± 0.02	0.30	0.69 ± 0.08	5
Palatinose	Glu α(1→6) Fru	0.03	--	0.49	--	4
Maltose	Glu α(1→4) Glu	0.01	0.24 ± 0.05	0.29	0.75 ± 0.06	5
Cellobiose	Glu β(1→4) Glu	0.01	--	0.33	--	5
Maltitol	Glu α(1→4) GluOH	0.03	--	0.34	--	5
Cellobiitol	Glu β(1→4) GluOH	0.03	--	0.50	--	5
Trehalose	Glu α(1→1) Glu	0.03	0.21 ± 0.03	0.26	0.83 ± 0.10	5
Lactulose	Gal β(1→4) Fru	0.01	0.18 ± 0.02	0.23	0.88 ± 0.08	5
β-Lactose	Gal β(1→4) Glu	0.01	--	0.31	--	5
Melibiose	Gal α(1→6) Glu	0.03	0.18 ± 0.03	0.37	0.68 ± 0.27	5
Lactitol	Gal β(1→4) GluOH	0.01	--	0.26	--	5
Melibiitol	Gal α(1→4) GluOH	0.03	--	0.23	--	5

* Abbreviations: Glu, glucose; Gal, galactose; Fru, fructose; GluOH, glucitol.
** Threshold is defined as the lowest concentration tested which elicited a measurable response in 50% of the animals.
CR_{50} = the concentration that evokes a response 50% of maximum
K_d = the dissociation constant
N = no. of animals
a = 95% confidence interval

Other D-Pyranosides. Further study of the taste
response to D-pyranose sugars revealed that the
replacement of the anomeric hydroxyl groups of α-D-
glucopyranose by a number of substituent groups led to
molecules of different stimulatory effectiveness
(Figure 5): OCH_3>F>OH>H. In addition to replacement
of a substituent group, the orientation of the OCH_3 or
OH group in the equatorial plane, as in methyl β-D-
glycopyranoside or β-D-glucopyranose, dramatically
reduced the effectiveness of the sugar as a stimulus
(Figure 6). A similar effect was observed for methyl
α-D-xylopyranoside and methyl β-D-xylopyranoside.

The compounds which differed from methyl α-D-
glucopyranoside by the replacement or reorientation of
the equatorial hydroxyl groups at positions C-2 or C-4
of the D-pyranoside ring were generally poorer stimuli
than methyl α-D-glucopyranoside. Methyl α-D-mannopyranoside, the C-2 axial epimer, and methyl α-D-galactopyranoside, the C-4 axial epimer, were considerably
poorer stimuli compared to the parent D-glucoside
(Figure 7). The response to the 2-deoxy derivative,
methyl 2-deoxy-α-D-arabino-hexopyranoside, was inferior
to the response of methyl α-D-glucopyranoside.

In contrast, stimulation with methyl α-D-xylopyranoside, which lacks a hydroxy-methyl group at the
C-5 position of methyl α-D-glucopyranoside, equaled the
response of methyl α-D-glucopyranoside (Figure 7A).

Polyols. Linear polyols containing two carbon to
seven carbon atoms all evoked neural responses. Two
important results were observed (Figure 8): (1) The
effectiveness ($CR_{50↓}$) of the polyols increased as the
chain length increased up to five carbon atoms; (2) In
contrast to monosaccharides, the configurations of
linear polyol may not play a role in the taste
response. This is indicated by the identical
responses to the four pentitols: D-arabinitol,
L-arabinitol, D-ribitol, or D-xylitol.

Competitive Interaction. The theoretical curve
for competitive interactions fits the actual data very
closely with the following discrepancies (Figure 9).
The mixture of 1.0 M methyl α-D-glucopyranoside and
sucrose and the mixture of 0.1 M methyl α-D-glucopyranoside and sucrose evoked responses slightly
greater and slightly lower, respectively, than predicted by the curve. In no case did the response to
the mixture ever exceed the maximum response evoked by
sucrose alone.

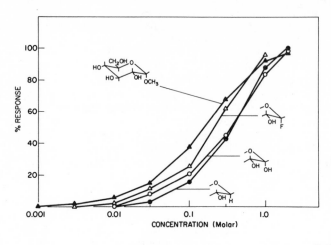

Brain Research

Figure 5. Effect of the substituent group at position C-1 on the stimulatory ability of D-*glucopyranose. Responses are relative to the maximum sucrose response. Methyl* α-D-*glucopyranoside* (▲), α-D-*glucopyranosyl fluoride* (△), α-D-*glucopyranose* (○), *1,5-anhydro-*D-*glucitol* (●) (13).

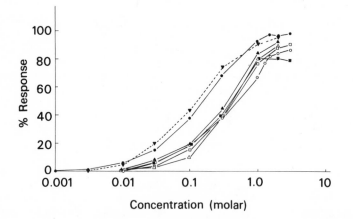

Figure 6. Effect of orientation of substituent group at C-1 on the stimulatory ability of a D-*glucopyranosyl derivative. Responses are relative to the maximum sucrose response. Methyl* α-D-*glucopyranoside* (●), *methyl* α-D-*xylopyranoside* (▼), α-D-*glucose* (▲), β-D-*glucose* (△) (*both of these sugars freshly made*), D-*glucose (equilibrium mixture)* (□), *methyl* β-D-*xylopyranoside* (■), *and methyl* β-D-*glucopyranoside* (○).

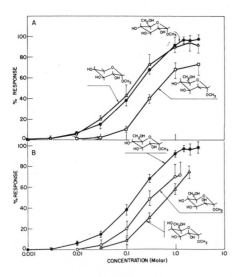

Brain Research

Figure 7. (A) Comparison of integrated chorda tympani nerve responses to methyl α-D-glucopyranoside (●) N = 15, methyl α-D-xylopyranoside (△) N = 5, and methyl 2-deoxy-α-D-arabino-hexopyranoside (☐) N = 5 solutions flowed over the tongue. Bars represent 95% confidence intervals. (B) Taste responses to methyl α-D-glucopyranoside (●) N = 15, methyl α-D-mannopyranoside (○) N = 6, and methyl α-D-galactopyranoside (△) N = 5. Responses relative to sucrose response of 100% (13).

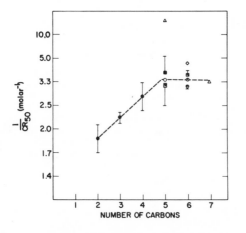

Figure 8. Relationship between the number of carbons in sugar alcohols and the reciprocal concentration which elicited a 50% response (CR_{50}). Bars indicate 95% confidence interval. Ethylene glycol (2C, ●, N = 5); glycerol (3C, ●, N = 5); erythritol (4C, ●, N = 6); D-ribitol (5C, ■, N = 6); L-arabinitol (5C, ▼, N = 6); D-arabinitol (5C, ○, N = 6); D-xylitol (5C, ☐, N = 5); D-sorbitol (6C, ☉, N = 5); D-galactitol* (6C, ◔, N = 5); D-mannitol* (6C, ◩, N = 5); myo-inositol* (6C, ◇, N = 10); perseitol* (7C, △, N = 6); sucrose (▽, N = 28). Asterisk (*) indicates sugars whose insolubility prevented direct determination of maximum response. The CR_{50} for these compounds was estimated from K_d.

124   CARBOHYDRATE–PROTEIN INTERACTION

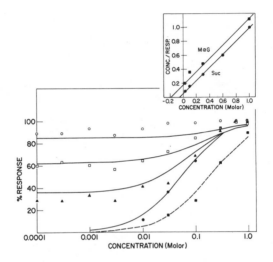

Brain Research

*Figure 9.   Concentration–response curve of sucrose in the presence of methyl
α-D-glucopyranoside (MαG). The solid lines are theoretical curves obtained from
an equation describing the competitive interaction of two substances with a single
receptor site (see below). Data points for sucrose alone (●); sucrose + 0.1 M
MαG (▲); sucrose + 0.3 M MαG, □; sucrose + 1.0 M MαG, ○; MαG alone (■).
Dashed line is the theoretical curve drawn from the binding equation for MαG.
K_d for sucrose (0.05 M) and K_d for MαG (0.18 M) were determined from the Beidler plot (see inset). CR_{50} for sucrose = 0.052 and for MαG = 0.18 (13).*

$$\left(\frac{Theoretical}{Response} = \frac{K_{suc}\,[M\alpha G] + K_{M\alpha G}\,[suc]}{K_{suc}\,[M\alpha G] + K_{M\alpha G}\,[suc] + K_{suc}\,K_{M\alpha G}} \right)$$

Similar to the mixtures of methyl α-D̲-gluco-
pyranoside and sucrose, mixtures of sucrose and
D̲-sorbitol closely fit the competitive interaction
curve (Figure 10). Response to the mixture gave a
less-than-additive effect at high concentrations. The
maximum response for the mixture did not exceed the
maximum response evoked by sucrose alone.

Discussion

As a taste stimulus, the effectiveness of sucrose
over other disaccharides and the effectiveness of
methyl β-D̲-fructofuranoside and methyl α-D̲-gluco-
pyranoside over other monosaccharides can be explained
by the presence of one or more of the following types
of sites: sucrose, α-D̲-glucopyranose and β-D̲-fructo-
furanose.

The β-D̲-Fructofuranose Site. In the gerbil MBFF
(methyl β-D̲-fructofuranoside) as well as sucrose
(α-D̲-glucopyranosyl - β-D̲-fructofuranoside)was the
most stimulatory D̲-fructose derivatives tested. This
finding suggests the presence of a specific fructose
site. This is not unlike the fly's D̲-fructose site
(23) which responds best to β-D̲-fructofuranose (24).
In contrast man may have a β-D̲-fructopyranose site
since this compound is believed to be the sweetest
fructose isomer (25,26). Further evidence for a
D̲-fructose site in other animals completely distinct
from a sucrose or a D̲-glucose site is evident in single
fiber responses (27,28) or in biochemical studies (29).
Some details of the specific D̲-fructose response
follow: (1) Those compounds such as 2-deoxy-D̲-fructose
(1,2-anhydro-D̲-mannitol) which lack a hydroxyl group or
contain a bulky substituent at C-3 (turanose), C-4
(lactulose) or C-6 (palatinose) are not as effective
stimuli as MBFF or sucrose (Sucrose can be considered
to be a D̲-fructose derivative with a bulky derivative
at C-2, i.e. α-D̲-glucopyranosyl-β-D̲-fructofuranoside).
The three D̲-fructose-containing disaccharides are
reducing sugars, unlike sucrose, and exist in solution
as a mixture of furanose and pyranose isomers. Should
the D̲-fructose site require the β-D̲-fructofuranose
ring form, sucrose would be the most stimulatory. (2)
The D̲-fructose derivatives which contain pyranoid rings
such as methyl α-D̲ and β-D̲-fructopyranosides were poor
stimuli compared to MBFF. (3) A furanoid ring D̲-fruc-
tose derivative in itself does not produce a good
response as indicated by the reduced response of methyl
α-D̲-fructofuranoside. (4) The equilibrated mixture of

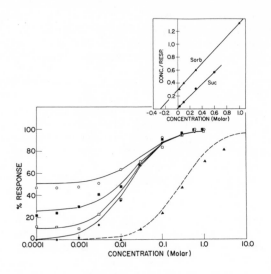

Brain Research

Figure 10.   Concentration–response curve of sucrose in the presence of D-sorbitol [sorb]. The solid lines are the theoretical curves obtained from the equation describing the competitive interaction of two substances with a single receptor site (see Figure 9). Data points for sucrose alone (●), sorbitol alone (▲), sucrose + 0.03 M [sorb] (□), sucrose + 0.1 M [sorb] (■), sucrose + 0.3 M [sorb] (○). Dashed line (– – –) is the theoretical curve drawn from the taste equation for sorbitol alone. K_d for sucrose = 0.015 M, CR_{50} for sucrose = 0.016 M, K_d for sorbitol = 0.29 M. Dissociation constants were determined from the Beidler plot (inset) (14).

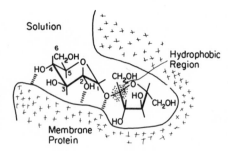

Brain Research

Figure 11.   Proposed model for the "sucrose site" in the membrane of the gerbil gustatory cell (14)

D-fructose (Figure 4), which has 31% β-D-fructo-
furanose (30), was also a highly stimulatory monosaccharide.

The α-D-Glucopyranose Site. Since methyl α-D-
glucopyranoside was the most stimulatory glucose
derivative tested, the presence of an α-D-glucopyranose
receptor site was suggested. This type of site has
been postulated in the blowfly and flesh fly (24,31).
Some of the characteristics of the taste response
to D-glucopyranosyl derivatives follow: (1) Unlike
the fly's taste receptor (24,31), there is no strict
configurational requirement for an α-glucoside in the
gerbil's taste response. Nevertheless α-D-glucopyranosyl derivatives produced larger responses than
the β-D-glucopyranosyl derivatives. In this respect,
the Mongolian gerbil compares favorably to humans (5,
16,20), the hamster (32) and other gerbil species (3)
in which α-D-glucopyranose derivatives are sweeter or
produce larger taste responses than β-D-glucopyranose
derivatives. (2) The requirement of equatorial
hydroxyls at C-2 and C-4 of the D-glucopyranoside
molecule for a maximum stimulation in the gerbil
parallels a similar requirement for increased sweet
taste in man (33) and maximum stimulation in the sugar
receptors of the gly (31). (3) The residues at the
C-5 position are less important in both the fly (31)
and gerbil. (4) The failure of the α-D-glucopyrano-
sides, turanose, palatinose, maltose, maltitol and
trehalose to stimulate as well as sucrose or methyl
α-D-glucopyranoside could be attributable to steric
hinderance involving the substituents at position C-1
of the glucopyranoside ring.

The Sucrose Site. The previous discussions cannot exclude the presence of a sucrose-specific site
since sucrose can fit into a D-glucose site and
D-fructose site simultaneously. A sucrose binding site
separate from a glucose or fructose binding site has
been found in cow tongue taste papillae (29). Based on
the following results the presence of a "sucrose site"
as shown in Figure 11 is suggested; (1) Among the
disaccharides sucrose was the most stimulatory sugar.
(2) Methyl α-D-glucopyranoside and methyl β-D-fructofuranoside, which have structural features in common
with sucrose were the most effective monosaccharides
for eliciting a neural response. (3) The lengths of
superimposed Dreiding models of a pentitol and sucrose
coincide almost perfectly. (This could account for the
leveling off of the response to the polyols as the

number of carbon atoms increase). (4) The responses
to mixtures of methyl α-D-glucopyranoside and sucrose
and mixtures of D-sorbitol and sucrose suggest that
these sugars act at a common receptor site.
    The "sucrose site" presented in Figure 11 would
evoke a response when a sugar molecule occupied it as
follows: The β-fructofuranosyl portion is tentatively
considered to occupy a "deep" subsite and the hydroxy-
methyl group at C-5 of the glucopyranoside is expected
to project into the solution. The deep subsite is
associated with a high degree of specificity as evidenced by the failure of the two fructosyl glucosides,
turanose (3-0-α-D-glucopyranosyl-D-fructose) and
palatinose (6-0-α-D-glucopyranosyl-D-fructose) to be as
stimulatory as sucrose. The monosaccharide response
data support the proposed binding of hydroxyl groups at
positions C-1, C-2, and C-4 of the D-glucose moiety.
The D-pyranosides which have equatorial substituents at
C-2 and C-4 and the C-1 axial substituent were the most
effective monosaccharides. A C-5 hydroxymethyl binding
locus is not required. This supports a model with the
C-6 hydroxy group protruding into the surrounding solution. The enhanced activity of methyl α-D-glucopyranoside over α-D-glucopyranose points to a hydrophobic region in the site. This coincides with the
relatively hydrophobic carbon atom of the fructose
moiety shown in Figure 11. Also, large bulky substituent groups at C-1 would block effective interaction
between the sugar and the sucrose site. Therefore,
this model would account for the relatively weak
responses of the other disaccharides as gustatory
stimulants.

Abstract

THE SPECIFICITY OF SUGAR TASTE-RESPONSE IN THE GERBIL

    Solutions of sugars were applied to the tongues of
gerbils, and the responses of their taste nerves
(chorda tympani) were measured electrophysiologically.
Sucrose elicited a nerve response at lower concentrations than any other substance tested. The ability of
a monosaccharide or polyol to elicit a nerve response
depends on the degree to which it resembles sucrose in
certain structural details; structures nearly identical to either half of the sucrose molecule give
responses at lower concentrations than molecules that
differ considerably from sucrose in conformation and
configuration. Responses to mixtures of sucrose with
methyl α-D-glucopyranoside or D-sorbitol indicate that

these sugars act at a common receptor site. These observations suggest the existence of a sucrose receptor site.

Literature Cited

1. Arnold, Walfred, N., J. Theoret. Biol., (1968) 21, 13.
2. Pazur, John H., In W. Pigman and D. Horton (Ed.), "The Carbohydrates", Academic Press, New York, (1970) pp. 69-137.
3. Jakinovich, William Jr. and Oakley, Bruce, J. Comp. Physiol., (1975) 99, 89.
4. Harper, Kenneth J., Kenagy, James G., and Oakley, Bruce, Neuroscience Abstracts, Society for Neuroscience (1976) 2 (part 1), 157.
5. Cameron, A. T., "The Taste Sense and the Relative Sweetness of Sugar and Other Sweet Substances". Sugar Research Foundation Report #9, (1947) pp. 1-74.
6. Moskowitz, Howard R., Amer. J. Psychol., (1971) 84, 387.
7. Kare, Morley. In L. M. Beidler (Ed.), "Handbook of Sensory Physiology", Vol. IV, Chemical Senses, Part 2. Taste, Springer-Verlag, Berlin, (1971) pp. 278-292.
8. Richter, Curt P., and Campbell, Kathryne H., J. Nutrition, (1940) 20, 31.
9. Hardiman, Clarence W., "Rat and Hamster Chemoreceptor Responses to a Large Number of Compounds and the Formulation of a Generalized Chemosensory Equation". PhD. Dissertation, Florida State Univ., (1964) Univ. Microfilms, Ann Arbor.
10. Hagstrom, E. C., and Pfaffmann, Carl, J. Comp. Physiol. Psychol., (1959) 52, 259.
11. Diamant, H., Funakoshi, M., Strom, L., and Zotterman, Yngve. In Y. Zotterman (Ed.), "Olfaction and Taste", Vol. I, Pergamon Press, Oxford, (1963), pp. 193-203.
12. Jakinovich, William, Jr., Brain Research, (1976) 110, 481.
13. Jakinovich, William, Jr., and Goldstein, Irwin J., Brain Research, (1976) 110, 491.
14. Jakinovich, William, Jr., and Oakley, Bruce, Brain Research, (1976) 110, 505.
15. Kimura, Katsumi and Beidler, Lloyd M., J. Cell. Comp. Physiol., (1961) 58, 131.
16. Cameron, A. T., Trans. Roy. Soc. Can., Sec. V., (1943) 37, 11.

17. Pangborn, Rose M. and Gee, Sandra C., Nature,
    (1961) 191, 810.
18. Pangborn, Rose M. and Chrisp, R. B., Experientia,
    (1966) 22, 612.
19. Tsuzuki, Yojiro, Kagaku (Tokyo), (1947) 17, 342.
20. Shallenberger, R. S. and Acree, T. E., In L. M.
    Beidler (Ed.), "Handbook of Sensory Physiology",
    Vol. IV, Chemical Senses, Part 2. Taste,
    Springer-Verlag, Berlin, (1971) pp. 221-277.
21. Steinhardt, Ralph G., Jr., Calvin, Allen D., and
    Dodd, Elizabeth A., Science, (1962) 135, 367.
22. Stewart, Roberta A., Carrico, Christine K.,
    Webster, Ronal L., and Steinhardt, Ralph G., Jr.,
    Nature, (1971) 234, 220.
23. Shimada, Ichino, Shiraishi, Akio, Kijima, Hiromasa,
    and Morita, Hiromichi, J. Insect Physiol., (1974)
    20, 605.
24. Hanomoni, Takamitsu, Shiraishi, Akio, Kijima,
    Hiromasa, and Morita, Hiromichi, Chem. Senses and
    Flavor, (1974) 1, 147.
25. Lindley, Michael G. and Birch, Gordon G., J. Sci.
    Fd. Agric., (1975) 26, 117.
26. Shallenberger, R., "Carbohydrates in Solution",
    Advances in Chemistry Series, Vol. 117, Amer.
    Chem. Soc., Washington, D. C., (1973) pp. 256-263.
27. Andersen, H. T., Funakoshi, M., and Zotterman,
    Yngve, Acta Physiol. Scand., (1962) 56, 362.
28. Pfaffmann, Carl, In J. T. Tapp (Ed.), "Reinforce-
    ment and Behavior", Academic Press, New York,
    (1969) pp. 215-241.
29. Lum, Clark K. and Henkin, Robert I., Biochim.
    Biophys. Acta, (1976) 421, 380.
30. Doddrel, David and Allerhand, Adam, J. Amer. Chem.
    Soc., (1971) 93, 2779.
31. Jakinovich, William, Jr., Goldstein, Irwin J., von
    Baumgarten, Rudolf J., and Agranoff, Bernard W.,
    Brain Research, (1971) 35, 369.
32. Noma, Akinori, Sato, Masayaso, and Tsuzuki, Yojiro,
    Comp. Biochem. Physiol., (1974) 48A, 249.
33. Shallenberger, R. S. and Acree, T. E., J. Agr.
    Food Chem., (1969) 17, 701.

RECEIVED September 8, 1978.

The Coordinated Action of the Two Glycogen Debranching Enzyme Activities on Phosphorylase Limit Dextrin

T. E. NELSON and R. C. WHITE

Department of Rehabilitation, Baylor College of Medicine, Houston, TX 77030

B. K. GILLARD

Department of Pediatrics, University of California School of Medicine, Los Angeles, CA 90024

Glycogen debranching enzyme (amylo-1,6-glucosidase/4-α-glucanotransferase) from rabbit muscle is the classical multi-catalytic site enzyme associated with phosphorylase which allows the total degradation of glycogen (1). The enzyme was discovered in 1950 by Cori and Larner when it was found that highly purified phosphorylase (1,4-α-D-glucan:orthophosphate α-glucosyltransferase, EC 2.4.1.1) would not completely degrade glycogen whereas less purified preparations would (2). They showed that phosphorylase could not go past the outer tier branch points of glycogen and produced a limit dextrin structure which contained the original branches with an average of four residues left on each chain. At the time it was thought that the limit dextrin produced by phosphorylase (∅-dextrin, LD) had an asymmetric structure with seven glucose units on the main chain and a single glucose unit remaining as the branch.

The debranching enzyme was thought to have a single activity, the function of which was to remove the glucose stub, thus allowing phosphorylase to further attack the debranched structure. The enzyme was called "amylo-1,6-glucosidase", in recognition of its action as a specialized glucosidase (2). About 10 years later it was found that the phosphorylase limit dextrin of glycogen (∅-dextrin) had a symmetrical outer tier structure with four residues on each chain and that amylo-1,6-glucosidase preparations contained a second activity. This second activity was a transferase which was suggested by Whelan and co-workers to disproportionate the branch chain by transferring a three unit segment to the main chain, elongating it and leaving a one unit stub (3,4). The presence of this second activity was confirmed by Cori, Illingworth and Brown and named "oligo-1,4→1,4-glucantransferase" (5,6). This sequence of actions and structures is shown in Figure 1. Subsequent work by Brown and Illingworth and their group, by Whelan and colleagues, and by Hers and co-workers showed that the two activities could be measured separately and independently of each other (1,7,8,9). The enzymatic activities can be measured by the original method of detecting glucose by

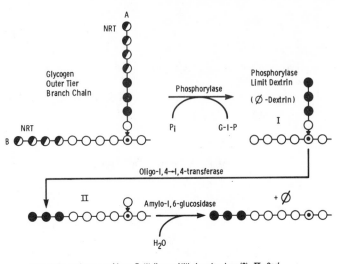

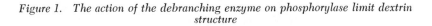

Legend: ○: α-glucose residues; I: Walker and Whelan structure (3); II: Cori and Larner structure (2); NRT: non-reducing terminal; A: A-chain; B: B-chain; ◐: glucose residues removed by phosphorylase; ●: glucose residues repositioned by the transferase;[—]: α-(I→4)-linkage; [↓]: α-(I→6)-linkage; ∅: free glucose residue.

Figure 1.   The action of the debranching enzyme on phosphorylase limit dextrin structure

Biochimica et Biophysica Acta

Figure 2.   Sodium dodecyl sulfate-poly-acrylamide gel electrophoresis of the purified debranching enzyme (26)

the combined action of both enzymes on limit dextrin which requires both activities (1,10,11). The glucosidase activity alone
can be measured by its ability to reincorporate glucose back into
polymer (10). This method was first developed by Hers and uti-
lizes ^{14}C-glucose (12,13). This action depends on the micro-
reversibility of the glucosidase reaction (14,15,16). The glucosidase action can also be measured by its ability to remove a
glucose stub from oligosaccharides such as the single unit
branched pentasaccharide "fast B$_5$" (6^3-α-glucosylmaltotetraose)
produced by α-amylase. This method was developed by Brown and
Illingworth (1,7,17). The glucosidase will also remove single
glucosyl stubs from singly branched α-glucosyl Schardinger dextrin
in an assay method developed by Whelan and his group (18). Brown
and colleagues have measured the transferase chromatographically
by its ability to disproportionate oligosaccharides (1,5,7,19).
In Larner's laboratory a kinetic method of measuring the transferase activity by its ability to disproportionate or attenuate
the outer chains of amylopectin was developed (20). The elongated outer chains have more amylose character and this can be
measured spectrophotometrically by the change in the iodine spectrum. This assay for transferase activity is independent of
glucosidase action (20). We have employed these three kinetic
methods of measuring the debranching enzyme system that we have
developed in the studies to be discussed: the combined activities (glucosidase-transferase) were measured by the production of
glucose from glycogen phosphorylase limit dextrin or "LD" (11).
The glucosidase activity alone was measured by ^{14}C-glucose incorporation into polymer (16) and the transferase activity alone was
measured by the change in the iodine spectrum of amylopectin (20).
    Subsequently the enzyme was purified in Larner's laboratory
to a significantly higher degree than previously reported. The
two activities were physically inseparable and resided in a homogeneous protein. Other groups also were unable to separate the
two activities. The molecular weight of the enzyme was thought to
be 260,000. Because of its large size, the enzyme was assumed to
be composed of subunits and to be double-headed; that is, two activities associated with a single homogeneous protein (7,23). The
possibility also remained that it was a multi-enzyme complex which
was difficult to separate into its component activities. Additional information pertaining to the debranching enzyme can be
found in several recent reviews (21,22). The fact remained that
all attempts to separate the enzyme into subunits or to separate
the activities from each other were unsuccessful (24,25).
    The reason for this is now clear. We have established that
the enzyme is a single polypeptide molecule under both denaturing
and non-denaturing conditions and that the molecular weight is
160-170,000 (26,27). The evidence for this is shown in Figure 2.
The figure shows the homogeneity of the protein using the methodology of Weber, Pringle and Osborn using reduction in the presence of 2-mercaptoethanol (MSH), alkylation by iodoacetic acid

134   CARBOHYDRATE–PROTEIN INTERACTION

(IAA) and denaturation in sodium dodecyl sulfate (SDS) (28).
Band A is the standard boiling water bath (BWB) method (Method 1).
Band B is omission of the BWB treatment, and C is an extension of
the BWB treatment. This indicates that no artifacts are produced
by contaminating proteases. Band D is an alternate control treatment (Method 2) where the enzyme is first denatured in 7 M
guanidine·HCl, then reduced in MSH followed by alkylation by IAA,
subsequent dialysis against 9 M urea and finally exposure to 0.1%
SDS. The latter treatment will dissociate any known protein into
component polypeptides (28). As can be seen the protein is a single polypeptide in all cases. Figure 3 shows the molecular weight
under denaturing conditions in SDS to be 160-170,000. Figure 4
shows the molecular weight to be the same under non-denaturing
conditions. The elution position of debranching enzyme activity
was detected directly without concentration so it is therefore
active as the 160,000 molecular weight species (27).
     The debranching enzyme (glucosidase-transferase) is therefore
active as a monomer and represents the first multi-catalytic site
enzyme of eucaryotic origin that functions as a single polypeptide
molecule (26,27,29). The debrancher is thus exceptional in this
regard. It is also unique in terms of the function that it per-
forms. As opposed to multi-enzyme complexes or multi-catalytic
enzymes, the debrancher does not catalyze two related chemical
reactions. Rather, it makes possible a discrete sequence of
changes in the physical structure of the substrate by two different reaction mechanisms. The structural changes involved result
in debranching. The two reactions have very different chemical
mechanisms. The first is an oligosaccharide transglycosylation;
an α-(1→4) bond of an oligosaccharide moiety is cleaved and then
reformed with another acceptor. There is no transfer to water
nor has the formation of another type of linkage ever been reported (21,22). This is apparently a strict disproportionation reaction catalyzed by the transferase. The reaction proceeds with no
change in free energy since both the linkage cleaved and the
linkage formed is α-(1→4). In this respect the transferase (1,4-
α-D-glucan:1,4-α-D-glucan 4-α-glycosyltransferase, EC 2.4.1.25)
resembles E. coli amylomaltase (1,4-α-D-glucan:D-glucose 4-gluco-
syltransferase, EC 2.4.1.3) and plant D-enzyme, both of which are
disproportionating transferases that reform the same type of
linkage they cleave (30). The second step, the hydrolysis of an
α-(1→6) linkage to form free glucose, is catalyzed by a specialized glucoside hydrolase. Although the free energy change in this
case is considerable, approximately 3-4000 cal., the glucosidase
is capable of catalyzing the reverse reaction, incorporation of
glucose into polymers such as glycogen, to a slight extent (10,
13,14,15,16,22,32). The enzyme will also transfer glucose to
other carbohydrate acceptors such as Schardinger dextrin, or maltotetraose to form an α-(1→6)-linked glucosyl Schardinger dextrin
or a branched pentasaccharide, or even to free glucose to form
isomaltose (15,33). In this respect the glucosidase (dextrin 6-

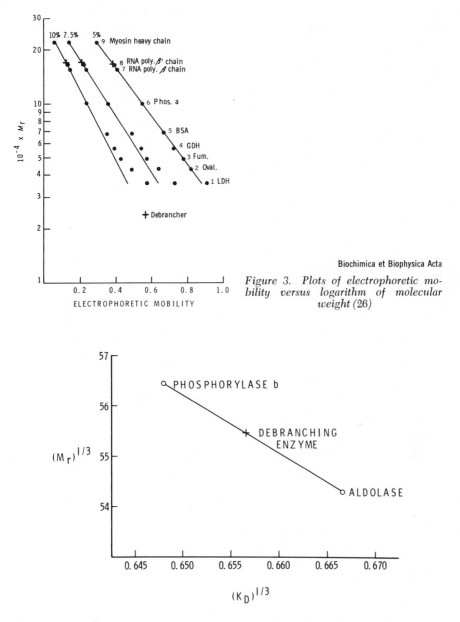

Biochimica et Biophysica Acta

Figure 3. Plots of electrophoretic mobility versus logarithm of molecular weight (26)

Biochimica et Biophysica Acta

Figure 4. Molecular weight determination by analytical gel chromatography under nondenaturing conditions. Plot of the cube root of the distribution coefficient versus the cube root of the molecular weight (27).

α-glucosidase, EC 3.2.1.33) is acting as a typical exo-glycosidase
that retains anomeric configuration in that it can catalyze either
hydrolysis or transglycosylation reactions (cf. ref. 31, p. 144
for a discussion of this relationship amongst exo-carbohydrases).
It is thus apparent that the catalytic reaction mechanisms of the
two activities are quite different.

Various studies have suggested that the transferase and glucosidase activities occur at distinct catalytic sites. The pH
optimum of the combined activities (glucosidase-transferase) on
limit dextrin is a function of the buffer. This is shown in
Figure 5. The pH optima is at 6.5 for anionic buffers but is
shifted to 7.2 by certain cationic buffers (11). Tris not only
shifts the pH optimum of the combined reaction but is inhibitory
as well. The pH optima of the two activities of the purified debrancher, when measured separately, are significantly different.
The pH optimum of the transferase (Figure 6) is at 6.0 (20) and
Tris has no effect. The glucosidase pH optimum (Figure 7) is at
6.5 (16) and here the effect of Tris can be seen. The reversible
inhibition by Tris of the glucosidase activity but not the transferase activity and the difference in their pH optima (20) suggest
that the two active sites have different catalytic groups.

Additional evidence for the separation of the two active
sites comes from measuring the two activities individually compared to their combined action on limit dextrin under mildly denaturing conditions (34). The effect on the two activities in the
presence of urea is shown in Figure 8. As can be seen the loss of
the combined activity is greater than the loss of the individual
activities in 2 M urea where the effect on the combined activity
is completely reversible. Figure 9 shows the effect of mild perturbation due to temperature when both the combined and the separate activities are measured. The inflection points in the
Arrhenius plots can be interpreted as evidence for a change in the
conformation of the polypeptide molecule. This data and the urea
data suggest an interrelationship between the glucosidase and
transferase active sites on the protein that can be affected in
their degree of coordination by conformational change. This suggests that the sites are separate and possibly occur in separate
domains of the folded polypeptide chain. Limited proteolysis
also indicates the existence of separate domains (35). Figure 10
shows the effect of limited proteolysis using trypsin. The large
130,000 molecular weight fragment and the small 35,000 mol. wt.
fragment appear initially and persist until all of the original
polypeptide disappears. This point (less than 1% debrancher remaining) in the course of the digestion is shown in the figure.
The same fragmentation pattern is also observed with both chymotrypsin and α-protease. This is shown in Table I. The same large
and small fragments are formed as seen with trypsin. This suggests that since the specificities of the three proteases are
quite different that they are all attacking a similar exposed area
of the folded polypeptide chain. This implies that the single

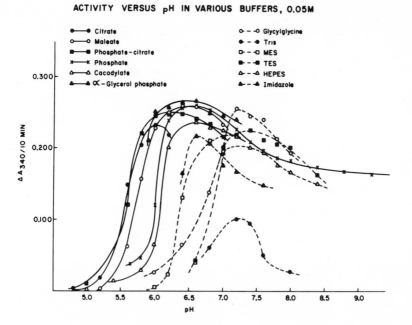

ACTIVITY VERSUS pH IN VARIOUS BUFFERS, 0.05M

Figure 5.   The hydrolysis of glycogen phosphorylase limit dextrin by the purified glucosidase-transferase as a function of pH in various buffers. Adapted from Nelson et al. (14).

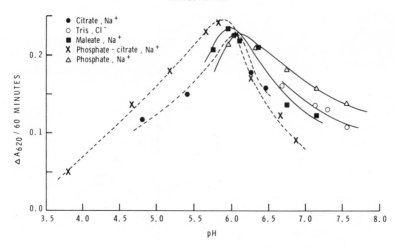

ACTIVITY VERSUS pH IN VARIOUS BUFFERS, 0.01M
AMYLOPECTIN

Biochimica et Biophysica Acta

Figure 6.   Activity of the transferase on amylopectin as a function of pH in various buffers (20)

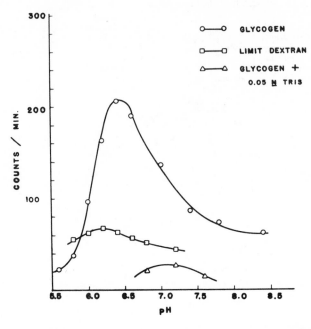

Analytical Chemistry

Figure 7.   Incorporation of glucose-¹⁴C into glycogen and glycogen phosphorylase limit dextrin as a function of pH in phosphate and Tris buffers (16)

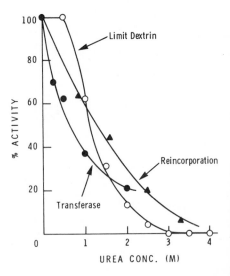

Molecular and Cellular Biochemistry

Figure 8.   The effect of urea concentration on the activities of the debranching enzyme as measured by different methods in the presence of urea (34)

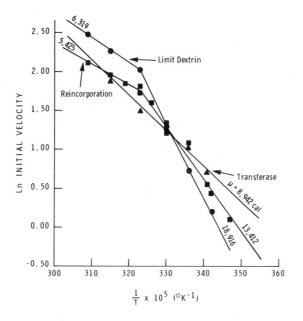

Molecular and Cellular Biochemistry

Figure 9. Arrhenius activation energy plots for the debranching enzyme as measured by different methods (34)

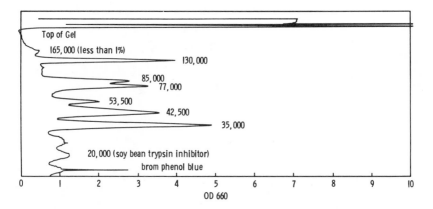

Figure 10. Sodium dodecyl sulfate-polyacrylamide gel electrophoresis of products of limited proteolysis of the debranching enzyme with trypsin. The molecular weights shown of the various bands were determined by the methodology described previously (26). The ratio of debrancher to trypsin was 100 to 1. The incubation was conducted for 60 minutes at 25°. The gel stain was Coomassie Brilliant Blue and the absorbance was measured at 600 nm using a Gilford gel scanner with a 0–1 O.D. chart scale (36).

TABLE I

Limited Proteolysis of Debranching Enzyme[a]

(Bands seen using SDS-P.A.G.E., M.Wt.)

Trypsin		Chymotrypsin		Alpha-Protease	
DB[b]:try, w/w, 100:1		DB:chym, w/w, 100:1		DB:α-prot, w/w, 100:1	
60 min, 25°		60 min, 25°		100 min, 25°	
lys, arg		tryp, tyr, phe, leu, met		leu, ileu, val	
165,000	very faint	165,000	major	165,000	major
130,000	major	130,000	minor	125,000	minor
		115,000	minor	105,000	faint
85,000	minor	86,000	minor	87,000	faint
77,000	minor	65,000	faint-minor	69,000	minor
53,000	faint	52,500	faint-minor	50,500	minor
42,500	minor	47,000	faint-minor	42,000	faint
35,000	major	39,000	faint		

[a] From White and Nelson (36).

[b] Debrancher

polypeptide chain is folded to comprise several domains and that the two activities may reside in separate domains since their active sites are different and their degree of combined action can be significantly affected by slight conformational changes.

Other investigators have also differentiated the two activities. The apparent absence of transferase but not glucosidase activity has been proposed as a subclass of type III glycogen storage disease (37). Exposure of the enzyme to guanidine inhibited glucosidase activity on "fast B_5" to a greater extent than the combined activity on "6^3-α-maltotriosylmaltotetraose" (B_7) (24) and partial proteolysis by exposure to trypsin or chymotrypsin destroys the combined activity to a greater extent than the glucosidase activity (38). These results also suggest that the transferase and the glucosidase activities are located at separate sites on the same enzyme molecule.

The debrancher thus represents a very unique situation of two mechanistically different activities that are associated with the same single polypeptide molecule, both of which are required to accomplish a particular function, namely, that of debranching the limit dextrin produced by phosphorylase. The question was how the two debrancher activities are coordinated.

To investigate this problem we chose to differentiate the two activities by use of substrate model inhibitors for the glucosidase. These inhibitors were used to distinguish the two activities with respect to their action on limit dextrin. In effect we undertook to "map" the active site of the enzyme kinetically (39). The rationale involved an attempt to explain the action of Tris as a non-competitive inhibitor (11) of the enzyme. Figure 11 shows this type of inhibition using limit dextrin as a substrate. The inhibition appears to be classically non-competitive. The next graph (Figure 12) is a plot of the slopes of the double-reciprocal plots versus inhibitor concentration. According to Cleland the fact that this is linear indicates that the inhibition is of the "simple linear" dead-end type (40). This means that the enzyme-inhibitor complex that forms is non-productive and that this applies to the enzyme-substrate-inhibitor complex (ESI) as well as to the enzyme-inhibitor complex (EI). This is represented in Equation 1a as shown. The general equation for inhibition is:

$$E + S + I \underset{K_m}{\overset{k}{\rightleftarrows}} ES + I \rightarrow \text{products} \qquad (1a)$$

$$K_{I_1} \updownarrow \qquad K_{I_2} \updownarrow$$
$$\text{EI} \qquad \text{ESI}$$

If $K_m + K_I$ steps are at equilibrium, then

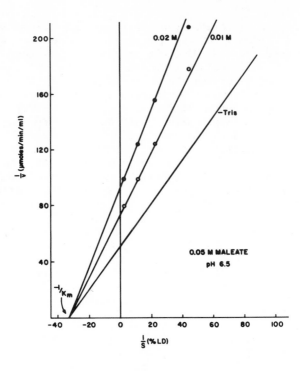

Biochemistry

Figure 11.   Double reciprocal plot of the action of the glucosidase-transferase on glycogen phosphorylase limit dextrin in the presence of Tris (11).

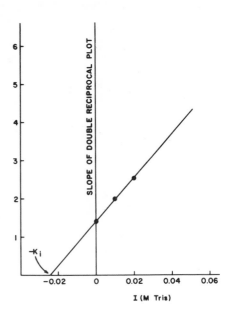

Figure 12.   Plot of the slopes of the double-reciprocal plots versus inhibitor (Tris) concentration. Adapted from data of Nelson et al. (11).

$$v = \cfrac{k[E]_T \, [S]}{K_m(1 + \cfrac{[I]}{K_{I_1}}) + [S](1 + \cfrac{[I]}{K_{I_2}})} \qquad (1b)$$

For non-competitive inhibition:

$K_{I_1} = K_{I_2} = K_I$, then

$$v = \cfrac{k[E]_T \, [S]}{(K_m + [S]) \, (1 + \cfrac{[I]}{K_I})} \qquad (1c)$$

Non-competitive inhibition involves the formation of both EI and ESI. The "simple linear" type indicates that they form with equal facility. This signifies that the dissociation constants for EI and ESI are the same (equation 1c).

A very important conclusion can be drawn from this result; namely that the enzyme can simultaneously bind inhibitor and substrate equally well. This implies that the binding site for inhibitor and the binding site for substrate are different. Thus, the binding site for Tris, the glucosidase inhibitor, is separate from the binding site for limit dextrin. Not only did this mean that the binding site for the transferase was separate from the glucosidase, since it was not inhibited by Tris, but also that there was possibly a third site for binding of polymer, i.e., a polymer binding site in addition to the two active sites.

How Tris affected the glucosidase site specifically and the relationship of the polymer binding site to the two active sites was investigated next. Several Tris analogs such as Bis-Tris were investigated and also found to be non-competitive inhibitors. Model building indicated that these hydroxyalkylamine compounds could mimic a portion of the structure of glucose if the amine nitrogen atom of the hydroxyalkylamine were juxtaposed at the position of the glucosidic ring oxygen. This is illustrated in Figure 13 using Bis-Tris as an example. The questions then became: could a structural relationship be observed, in terms of mimicking a glucose molecule, and if so, was there a correlation with the degree of charge, i.e., was a protonated amine required? The series of compounds we chose to examine this relationship is given in Table II. The compounds are drawn to mimic the structure of glucose. Nojirimycin is similar in structure to glucose except for a nitrogen in place of the ring oxygen. It is the best inhibitor. The acyclic hydroxyalkylamines are poorer, by one to three orders of magnitude. As can be seen, erythritol, which has no charged amino atom, is a very poor inhibitor. Threitol is an

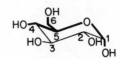

α-D-GLUCOPYRANOSE

Figure 13. Structures of α-D-glucopyranose and Bis-Tris showing superimposition at C_4, C_5, C_6, and C_1 of glucose. Glucose is shown in the C-1 chain conformation. The heavy lines indicate the portions of the structures that are superimposable.

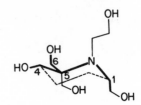

BIS-TRIS

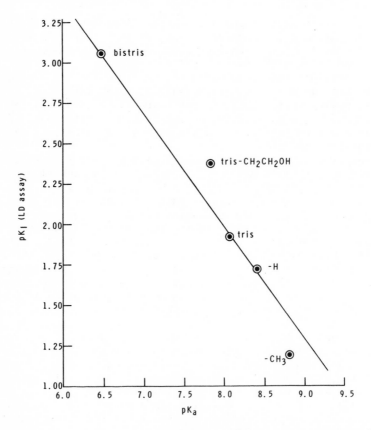

Figure 14. Plot of pK_I of hydroxylalkylamine inhibitors versus pK_a. Numbers correspond to compounds 2, 3, 5, 6, and 7 shown in Table II. Adapted from Gillard and Nelson (39).

even poorer analog of glucose than erythritol, since it can mimic
even less of the ring hetero atom region of glucose, i.e., the
hydroxyalkyl substituents about C_4, C_5, C_6 and C_1. Threitol is
in fact not detectable as an inhibitor at the concentration shown.
Likewise, the cyclic sulfur analogs are also poor inhibitors due
to the lack of charge on the sulfur atom. The acyclic sulfur compounds or acyclic alkyl nitrogen compounds that lack hydroxy functions are also very poor or non-inhibitors.
     These data indicate that not only do hydroxyalkyl groups pro-
mote inhibition but that a charged amino atom enhances it still
further. With regard to the acyclic amines there is a direct cor-
respondence between the pK of the amine and its degree of inhibition. This is shown in Figure 14. Within a given class of compounds the degree of inhibition is related to the ability of the
inhibitor to donate a proton. To verify that these inhibitors
were actually acting at the glucosidase site we reacted the debrancher with an active-site-directed irreversible inhibitor, 1-
S-dimethylarseno-1-β-D-glucoside in the presence of the reversible
inhibitor, Bis-Tris (Fig. 15). The rate of loss of glucosidase
activity is decreased in the presence of Bis-Tris, indicating that
the two inhibitors bind at the same or interacting sites (39,41).
     Further indication that the low molecular weight inhibitors
(Table II) are acting specifically at the glucosidase site comes
from the fact that they are non-competitive when measured by the
combined reaction, but are competitive when measured by the glucosidase, glucose-reincorporation reaction. D-glucono-1,5-lac-
tone is also an inhibitor of the glucosidase with a K_I of 3.2 mM,
or approximately in the same range as the acyclic hydroxyalkylamines (39). Thus these inhibitors of the glucosidase are acting
like charged glucosyl cations. They appear to be mimicking an
activated form of glucose formed by the glucosidase during the
course of hydrolysis, and are thus activated or "transition state"
analogs rather than ground state analogs such as glucose, whose
K_m is 32 mM or an order of magnitude less effective in binding
than Bis-Tris.
     The hydrolysis of glycosides by mineral acid is
thought to proceed via a glucosyl carbonium-oxonium ion as shown
in Figure 16 (42). It has been postulated that carbohydrases
that retain anomeric configuration proceed via a glycosyl-enzyme
intermediate (43). We have shown previously that the glucosidase
forms a glucosyl-enzyme intermediate; that is, it will transfer
glucose in the forward and reverse directions. The enzyme also
retains the initial anomeric configuration of the substrate in
the product (14,15). The formation of the α-anomer is demonstrated by the mutarotation shown in Figure 17. It is thus possible that the mechanism of hydrolysis proceeds via a charged
intermediate with cationic character. The similarities between a
glucosyl carbonium-oxonium ion structure and that of gluconolactone and Nojirimycin are illustrated in Figure 18. To further
investigate whether such a charged structure occurred, a Brønsted

INHIBITION OF DEBRANCHING ENZYME

No.	Name	Structure [a]	pK_a [j]	Combined Activity [b] K_i, mM	Glucosidase Activity [c] K_i, mM
1.	Nojirimycin, 5-amino-D-glucose	HO, N, OH, ~OH, HO, OH	5.3	0.024 ± 0.004	0.0039 ± 0.0005
2.	Bistris	HO, N, OH, OH, OH, HO	6.48	1.66 ± 0.05 2.0 ± 0.2 [e]	1.39 ± 0.20 [d]
3.	Hydroxyethyl tris	HO, N, OH, HO, OH	7.83	4.2 ± 0.6 [f]	1.7 ± 0.5
4.	Bis(tris) propane	HO, N, N, OH, OH, HO, OH, HO	~6.9	14.8 ± 0.8	—
5.	Tris	HO, N, OH, HO	8.08	11.8 ± 0.7 [f] 23 ± 1 [g]	6.3 ± 1.0
6.	2-amino-1,3-propanediol	HO, N, HO	8.4	19 ± 6 [f]	3.8 ± 1.0
7.	2-amino-2-methyl-1,3-propanediol	HO, N, HO	8.80	64 ± 10 [f]	27 ± 6
8.	m-Erythritol	HO, OH, OH, HO		$154 + 11$	$86 + 8$
9.	D,L-threitol	OH, OH, OH, HO, OH, OH, OH, HO		no inhibition (100 mM) [h]	no inhibition (250 mM) [h]
10.	5-thio-D-glucose	HO, S, ~OH, OH, HO, OH		~243 [i] (25 mM) [h]	—
11.	cyclic DTE	S, S, HO, OH		~150 [i] (12.5 mM) [h]	no inhibition (31 mM) [h]
12.	cyclic DTT	S, S, OH, HO		~370 [i] (25 mM) [h]	no inhibition (63 mM) [h]

No.	Name	Structure[a]	pK_a[j]	Combined Activity[b] K_i, mM	Glucosidase Activity[c] K_i, mM
13.	DTE			no inhibition (100 mM)[h]	no inhibition (250 mM)[h]
14.	DTT			no inhibition (100 mM)[h]	no inhibition (250 mM)[h]
15.	Putrescine		9.35[k] 10.80	~300[i] (50 mM)[h]	—
16.	Cadaverine		9.74[k] 11.05	no inhibition (50 mM)[h]	—
17.	Spermidine			no inhibition (50 mM)[h]	—
18.	Spermine			no inhibition (50 mM)[h]	—
19.	DAPH			~220[i] (50 mM)[h]	—
20.	HEPA			~120[i] (50 mM)[h]	—
21.	α-Schardinger dextrin	$\left[\text{Glucosyl}\right]_6$ $\alpha\text{-}(1\rightarrow 4)$		0.76 ± 0.05[e] mg/ml	—
22.	Glycogen	$\alpha\text{-}(1\rightarrow 4)$-linked glucose polymer with $\alpha\text{-}(1\rightarrow 6)$- branch points		0.56 + 0.14 mg/ml	—

[a] Drawn to mimic glucose structure.
[b] Rates measured using the standard assay method, unless noted otherwise.
[c] Rates measured using the [¹⁴C] glucose incorporation assay. Except for bistris, K_i values calculated from Dixon plots, assuming competitive inhibition.
[d] Calculated by method of Eisenthal and Cornish–Bowden (1974).
[e] Rates measured using coupled enzyme assay.
[f] Calculated from Dixon plot assuming noncompetitive inhibition.
[g] Value of Nelson et al. (11).
[h] Maximum concentration tested.
[i] Calculated from relative activity in presence and absence of noted inhibitor concentration, assuming noncompetitive inhibition.
[j] For dissociation of conjugate acid of amines.
[k] Robinson and Stokes (46).

Adapted from Gillard and Nelson (39).

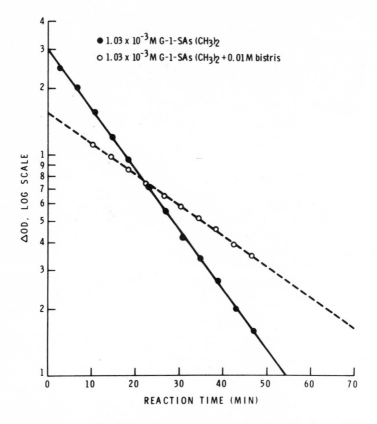

Figure 15. Rate of the debranching enzyme on phosphorylase limit dextrin in the presence of the irreversible inhibitor dimethylarsenothioglucose in the presence and absence of a reversible inhibitor (Bis-Tris). Adapted from Gillard et al. (41).

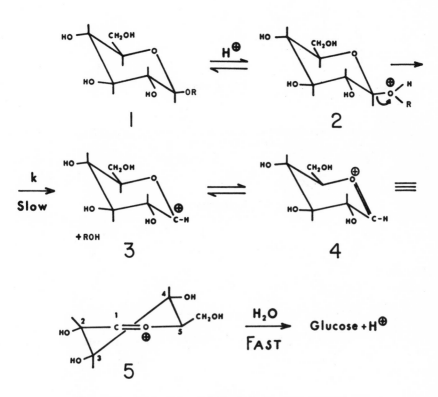

Figure 16. *Acid-catalyzed hydrolysis of a glycosidic linkage*

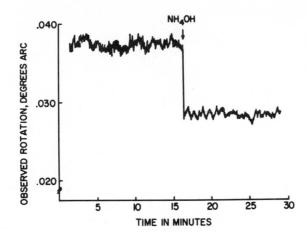

Biochimica et Biophysica Acta

Figure 17.   Observed optical rotation during glucosidase–transferase hydrolysis
of phosphorylase limit dextrin (15)

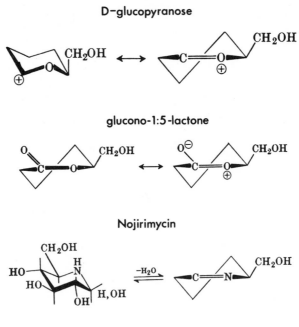

Biochemistry

Figure 18.   Similarities in structure between a glucosyl carbonium-oxonium ion
and gluconolactone and Nojirimycin

plot of the amine inhibitors was constructed. This is shown in
Figure 19 for both the combined reaction and the glucosidase
alone. Plots A and B are based on the total inhibitor concentration and C and D are based on only the concentration of the
protonated species. As can be seen the plots are all essentially
similar. Plot D is the glucosidase plot using the protonated inhibitor concentration. The slope here is approximately 1.3. The
Brønsted relationship indicates whether or not proton donation
takes place between enzyme and inhibitor and the direction in
which it occurs. The positive value of 1.3 indicates that proton
donation does occur and that it occurs from inhibitor to enzyme
(39). This indicates that a positively charged ionic form of
glucose may occur during the course of hydrolysis and that there
is a corresponding nucleophilic group at the catalytic site of
the glucosidase (39). Thermodynamic calculations made from the
dissociation and equilibrium constants involved indicate that the
pK of the group is approximately 8.5 (39). This implicates
cysteine. Inhibition with sulphydryl reagents such as PCMB and
Ellman's reagent (DTNB) and water soluble alkyl carbodiimide (EDC)
suggest that both an active SH group and a carboxylate anion are
present at the active site (44). This information combined with
the structural properties of the hydroxyalkylamines in terms of
their comparative effectiveness as inhibitors suggests a hypothetical representation for the glucosyl ion in the glucosidase
active site (Figure 20). From the amine inhibition data, the
glucosyl C_3 region seems to be uninvolved in the structural specificity. It is of interest to note that if this region of the active site were an unencumbered area it would allow water as a
nucleophile to enter from below the glucosyl carbonium-oxonium
ion structure in the conformation shown. This would result in
the formation of an asymmetric center having the α-anomeric configuration as is found experimentally.

In summary, all of the low molecular weight inhibitors shown
in Table II are non-competitive when measured by the combined
reaction and the evidence indicates this is due to their being
able to act as "transition state analogs" at the glucosidase site
by "mimicking" an activated structure of glucose. To confirm this
contention their inhibitory action was measured using the ^{14}C-
glucose-glucosidase reincorporation reaction. In this case the
amine inhibitors (cf. Table II) were competitive, as shown with
Bis-Tris in Figure 21. Similarly, one would predict, when
measuring the combined reaction, that if the polymeric binding
site for substrate were separate from the glucosidase binding
site then the binding of polysaccharide inhibitors should be competitive. Using glycogen, as shown in Figure 22, the inhibition
is competitive. Glycogen resembles debranched limit dextrin and
is a very poor substrate. Competitive inhibition was also found
with α-Schardinger dextrin (39). The implication here is clear.
The competitive inhibition indicates that there are not two
separate binding sites for polysaccharide but rather that the

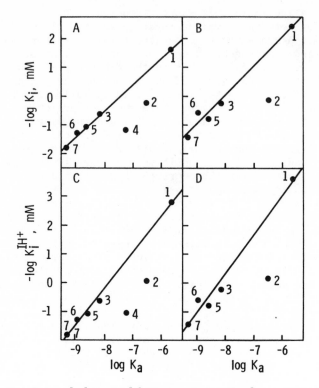

Figure 19. Brønsted plots of inhibition constants. Numbers refer to inhibitors
in Table II (39).

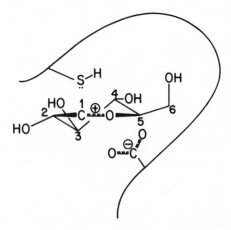

Figure 20.   Representation of a glycosyl
ion in the glucosidase active site

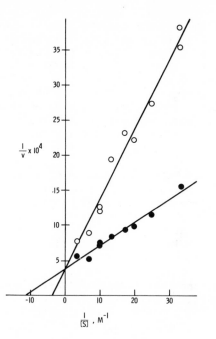

Biochemistry

Figure 21. Double-reciprocal plot of Bis-Tris inhibition of debranching enzyme glucosidase activity using glucose-^{14}C assay (39)

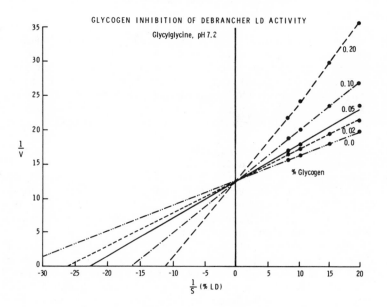

Figure 22. Double-reciprocal plot of glycogen inhibition of debranching enzyme activity on phosphorylase limit dextrin. Adapted from Gillard and Nelson (39).

polysaccharide substrate site and the polymer binding site are the same. Although the competitive inhibition clearly indicates that there are not two separate sites for polymer, it does not indicate whether there is a single site or two strongly interacting or over-lapping sites. The kinetic evidence obtained at this point cannot distinguish between these latter alternatives. The fact that the glucosidase incorporates free glucose into polymers indicates that the polymer site is not exclusively concerned with transferase action. It should be noted that, as in the case of Tris (20), none of the small molecular weight inhibitors had any demonstrable effect on the transferase activity (39).

The evidence presented indicates that the debranching enzyme has three separate portions to its active site: a transferase active site; a glucosidase active site; and a polymer binding site. The polymer binding site is probably a series of adjacent glucosyl residue subsites that are shared by both activities and represents an aglycone or main chain binding site. It is pos-sible (though hypothetical at this time) that the binding site portion of the transferase active site also consists of a series of three glucosyl residue subsites and that the glucosidase bind-ing site likewise consists of a single subsite, and that these are contiguous with the subsites of the polymer binding site. The transferase active site and the glucosidase active site thus bind their respective glycone portions of the branch chain (A-chain) and the polymer site binds the remaining or aglycone por-tion of the substrate. This would be the main chain (B-chain) backbone that is not altered.

A possible mechanism for debrancher action on phosphorylase limit dextrin is shown in Figure 23. Here the two activities act in a concerted or cooperative manner on a single outer tier branch. The transferase site binds the maltotriosyl portion of the branch chain (A-chain) and transfers the reducing terminal of the third residue to the non-reducing terminal of the main chain (B-chain). The exposed 1,6-linked glucose residue is then free to shift about the branch linkage to align itself on the nearby but separate glucosidase site for hydrolysis. Whether completed transfer and hydrolysis occur simultaneously or sequen-tially is not indicated in the model, since the present results cannot distinguish between these possibilities or alternatives. It is possible that the structural conformational change produced in transfer of the maltotriosyl moiety induces glucosidase action. The effect of mildly denaturing conditions on the coordination of the two activities might suggest this.

A space filling model of the substrate indicates that this mechanism is dimensionally plausible. If the outer tier chains are placed in a left-handed helical conformation as diagrammed in Figure 24, then the third residue of the branch chain is in close proximity to the terminal residue of the main chain so that transfer is easily possible. The sequence of enzyme catalyzed structural alterations is shown using space filling (CPK)

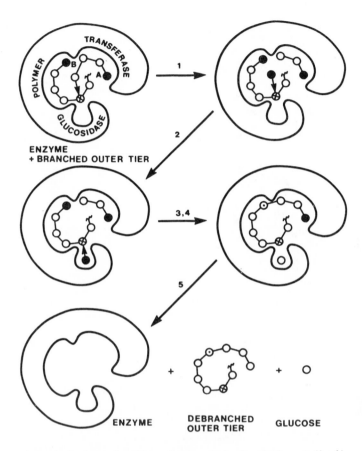

Legend: ○ glucosyl residue; ● nonreducing terminal; — α-(1→4)-linkage; ↓ α-(1→6)-linkage. Proposed mechanism: 1) Transferase removes maltotriosyl group; 2) Glucosyl residue shifts to glucosidase site; 3) & 4) Maltotriosyl group transferred to nonreducing terminal and (1→6)-linkage cleaved; 5) Release of products.

Biochemistry

Figure 23. *Proposed action of debrancher on phosphorylase limit dextrin (39)*

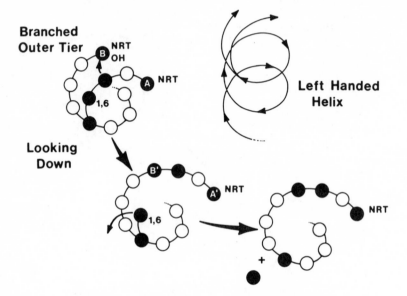

Figure 24. Dimensional representation of phosphorylase limit dextrin structure and changes produced by action of the debranching enzyme

molecular models. As can be seen in Figure 25, the third glucosyl residue from the branch chain non-reducing terminal is located so that a maltotriosyl residue can easily shift slightly to
attach to the main chain non-reducing terminal residue free C_4
hydroxyl group. Figure 26 shows this transfer without movement
of the main chain. The remaining glucose stub, now exposed, can
rotate slightly as seen in Figure 27 (shown as rotated ca. 90°
outward, or clockwise), and can thus realign itself in a sector
on the enzyme different from its original orientation, i.e., on
an adjacent though separate glucosidase site for hydrolysis.
Figure 28 shows removal of the glucose stub and the resultant
debranched outer chain. It is of interest that all of the active
sites proposed in the mechanism for the enzyme could occur in a
concave fold or cleft region on the protein that could fit over
the top, around and somewhat below the outside rim of the outer
tier of the molecule as shown in Figure 25. This arrangement of
active sites would permit the catalytic groups of the transferase
site to interact with the glucosidic linkage of the third residue
of the A-chain and yet not conflict with the catalytic groups of
the proposed adjacent glucosidase site. It must be emphasized
that no definitive conclusions on this arrangement of sites are
possible at this time. The postulated conformation of the substrate still allows many possible alternative arrangements and
interplay. The proposed mechanism, however, is physically
plausible in terms of the three dimensional structure of the substrate as shown.

As postulated in Figure 28, a linear dextrin chain can be
accommodated by the transferase site and the polymer or backbone
site without the involvement of the glucosidase site. Evidence
in the literature indicates that the transferase action is reversible, both intramolecular transfer as shown here with phosphorylase limit dextrin, and intermolecular transfer as evidenced with
branched and linear oligosaccharides (1,5,7,13,19,20). Thus a
linear dextran chain, from amylopectin or glycogen, could bind to
the transferase and polymer site and a maltotriosyl enzyme intermediate could be formed. If the aglycone portion of the chain
were to diffuse away and be replaced on the polymer site by
another linear chain whose non-reducing terminal could act as an
acceptor for the maltotriosyl group, one would have reattachment
or transfer. Such elongation or attenuation of chains has been
found experimentally (6,13,10,20,45). The proposed model for the
debrancher active site could accommodate this type of action on
linear chains by the transferase as readily as it can accommodate
glucose incorporation into polymer by the glucosidase.

The coordinated mechanism proposed here for debrancher action
on phosphorylase limit dextrin, although hypothetical, is consistent with the evidence that the debrancher has two distinct
catalytic sites and a polymer binding site or sites that overlap
or interact strongly. It accounts for the observed inhibition
patterns with limit dextrin and glucose incorporation and simul-

Figure 25. Molecular model of glycogen phosphorylase limit dextrin outer tier branch chain before debrancher action. Space filling CPK models were used. The nonreducing terminal free C_1-OH group of the B-chain which accepts the reducing terminal of the A-chain maltotriosyl residue is indicated by arrow 1. The glucosidic linkage oxygen of the maltotriosyl group reducing terminal is indicated by arrow 2. The glucosidic oxygen of the α-$(1 \rightarrow 6)$ branch chain linkage is shown by arrow 3.

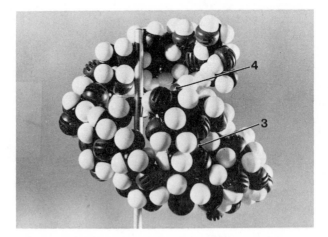

Figure 26. Transfer of maltotriosyl residue from A-chain to B-chain. This rearrangement corresponds to stage 2 in Figure 24. The exposed glucosyl stub left on the A-chain in the foreground. The free C_4OH of the stub produced by the transfer is shown by arrow 4. The branch chain linkage is shown as before by arrow 3.

Figure 27.   Rotation of exposed glucosyl branch chain residue to orientation on the glucosidase site. The rotated A-chain stub is in the foreground. The free C_1-OH of the stub is again shown by arrow 4. The glucosidic oxygen of the linkage formed by the transfer of the maltotriosyl group to the main chain is shown by arrow 5. (This linkage is not visible in Figure 26.) The glucosidic oxygen of the α-(1 → 6) linkage of the branch chain stub shown in Figures 25 and 26 is not visible in this view.

Figure 28.   Debranched outer tier of phosphorylase limit dextrin by action of debranching enzyme. The elongated main chain (B-chain) is shown after removal of the A-chain glucosyl stub. This corresponds to stage 3 in Figure 24. The C_6-OH group of the main chain residue that formed the branch chain linkage is shown by arrow 6.

taneous binding of glucosidase inhibitors and polysaccharides.
It is also consistent with transferase action and reverse glucosidase action occurring independently of each other via a shared
polymer site. It will be of considerable interest to see where
further work with this unique mammalian enzyme will lead and
whether it will substantiate the model proposed here on the basis
of the present results.

Abstract

Rabbit muscle glycogen debranching enzyme (amylo-1,6-gluco-
sidase/4-α-glucanotransferase) is a single polypeptide chain
(mol. wt. 160-170,000) which has two distinct but functionally
interrelated activities. It is the first such bi-functional
eucaryotic enzyme to be reported that is active as a monomer.
Limited proteolysis indicates that the two catalytic sites may
reside in different domains of the molecule and that their interaction is affected by conformational changes in the protein. The
evidence indicates that although their catalytic sites are
separate, the transferase and glucosidase each have a glycone
binding site for their respective portions of the substrate and
a common aglycone binding site for polysaccharide residues. The
relationship between the catalytic and binding sites has been
investigated using reversible substrate model inhibitors. A
Brønsted plot indicates that a proton transfer occurs from inhibitor to enzyme during the binding process and that the inhibitors may mimic a glucosyl ion structure, suggesting that this
species may be an intermediate in the enzyme catalyzed hydrolysis.
A mechanism for the cooperative behavior of the two activities
suggests a coordinated action on the part of the enzyme to
facilitate debranching of the limit dextrin structure.

Literature Cited

1. Brown, D.H., and Brown, B.I., Methods Enzymol. (1966) 8,
    515-524.
2. Cori, G.T., and Larner, J., J. Biol. Chem. (1951) 188, 17-21.
3. Walker, G.J., and Whelan, W.J., Biochem. J. (1960) 76, 264-
    268.
4. Abdullah, M., and Whelan, W.J., Nature (1963) 197, 979-980.
5. Brown, D.H., and Illingworth, B.I., Proc. Natl. Acad. Sci.
    U.S. (1962) 48, 1783-1787.
6. Brown, D.H., Illingworth, B., and Cori, C.F., Nature (1963)
    197, 980-982.
7. Brown, D.H., and Illingworth, B., in "Control of Glycogen
    Metabolism", pp. 139-150, Whelan, W.J., and Cameron, M.P.,
    Eds. Churchill. London 1964.
8. Abdullah, M., Taylor, P.M., and Whelan, W.J., in "Control of
    Glycogen Metabolism", pp. 123-138, Whelan, W.J., and
    Cameron, M.P., Eds. Churchill. London 1964.

9. Hers, H.G., Verhue, W., and Mathieu, M.,, in "Control of Glycogen Metabolism", pp. 151-163, Whelan, W.J., and Cameron, M.P., Eds. Churchill. London 1964.

10. Larner, J., and Schliselfeld, L.H., Biochim. Biophys. Acta (1956) 20, 53-61.

11. Nelson, T.E., Kolb, E., and Larner, J., Biochemistry (1969) 8, 1419-1428.

12. Hers, H.G., Rev. Intern. Hepatol. (1959) 9, 35-55.

13. Hers, H.G., Verhue, W., and Van Hoof, F., European J. Biochem. (1967) 2, 257-264.

14. Nelson, T.E., Kolb, E., and Larner, J., Biochim. Biophys. Acta (1968) 151, 212-215.

15. Nelson, T.E., and Larner, J., Biochim. Biophys. Acta (1970) 198, 538-545.

16. Nelson, T.E., and Larner, J., Anal. Biochem. (1970) 33, 87-101.

17. Illingworth, B., and Brown, D.H., Proc. Natl. Acad. Sci. U.S. (1962) 48, 1619-1623.

18. Taylor, P.M., and Whelan, W.J., Arch. Biochem. Biophys. (1966) 113, 500-502.

19. Brown, D.H., and Brown, B.I., Abstracts, 154th Meeting of the American Chemical Society, 1967, p. 54D.

20. Nelson, T.E., Palmer, D.H., and Larner, J., Biochim. Biophys. Acta (1970) 212, 269-280.

21. Ryman, B.E., and Whelan, W.J., Advan. Enzymol. (1971) 34, 285-443.

22. Lee, E.Y.C., and Whelan, W.J., Enzymes, 3rd Ed. (1971) 5, 191-234.

23. Cori, C.F., in "Control of Glycogen Metabolism", p. 169, Whelan, W.J., and Cameron, M.P., Eds. Churchill. London 1964.

24. Brown, D.H., Gordon, R.B., and Brown, B.I., Ann. N.Y. Acad. Sci. (1973) 210, 238-253.

25. Lee, E.Y.C., and Carter, J.H., FEBS Lett. (1973) 32, 78-80.

26. White, R.C., and Nelson, T.E., Biochim. Biophys. Acta (1974) 365, 274-280.

27. White, R.C., and Nelson, T.E., Biochim. Biophys. Acta (1975) 400, 154-161.

28. Weber, K., Pringle, J.R., and Osborn, M., Methods Enzymol. (1972) 26, 3-27.

29. Taylor, C., Cox, A., Kernohan, J., and Cohen, P., Eur. J. Biochem. (1975) 51, 515-524.

30. French, D., in "Biochemistry of Carbohydrates", pp. 267-335, Whelan, W.J., Ed. University Park Press. Baltimore, Md. 1975.

31. Nelson, T.E., Biochim. Biophys. Acta (1975) 377, 139-145.

32. Lee, E.Y.C., and Carter, J.H., Arch. Biochem. Biophys. (1973) 154, 636-641.

33. Stark, J.R., and Thambyrajah, U., Biochem. J. (1970) 120, 17-18 p.

34. Nelson, T.E., and Watts, T.E., Mol. Cell. Biochem. (1974) 5, 153-159.
35. Kirschner, K., and Bisswanger, H., Ann. Rev. Biochem. (1976) 45, 143-166.
36. White, R.C., and Nelson, T.E., unpublished results.
37. Van Hoof, F., and Hers, H.G., Eur. J. Biochem. (1967) 2, 265-270.
38. Bates, E.J., Heaton, G.M., Taylor, C., Kernohan, J.C., and Cohen, P., FEBS Lett. (1975) 58, 181-185.
39. Gillard, B.K., and Nelson, T.E., Biochemistry (1977) 16, 3978-3987.
40. Cleland, W.W., Biochim. Biophys. Acta (1963) 67, 104-137; 173-187; 188-196.
41. Gillard, B.K., Zingaro, R.A., and Nelson, T.E., Abstracts, 29th Southwest Regional Meeting of the American Chemical Society, 1973, p. 37.
42. Overend, W.G., in "The Carbohydrates", pp. 279-353, Pigman, W., and Horton, D., Eds. Second Ed. Academic Press, N.Y., Vol. 1A, 1972.
43. Koshland, D.E., Jr., Enzymes, 2nd Ed. (1959) 1, 305-346.
44. Nelson, T.E., Gillard, B.K., and White, R.C., unpublished results.
45. Brown, D.H., Brown, B.I., and Cori, C.F., Arch. Biochem. Biophys. (1966) 116, 479-486.
46. Robinson, R.A., and Stokes, R.H.,"Electrolyte Solutions", p. 526, 2nd Ed. Rev. Butterworths. London 1968.

Acknowledgements

This work was supported in part by grants from The Robert A. Welch Foundation (Q-402) and the National Institutes of Health (AM-13950, AM-17978 and 1-RO2-AM-57255).

RECEIVED September 8, 1978.

The Role of Glycosidically-Bound Mannose in the Cellular Assimilation of β-D-Galactosidase

JACK DISTLER, VIRGINIA HIEBER, ROY SCHMICKEL, and GEORGE W. JOURDIAN

The Rackham Arthritis Research Unit and The Departments of Biological Chemistry and Pediatrics, The University of Michigan, Ann Arbor, MI 48109

While glycose-protein complexes have been recognized since the early 1800's (1,2), the role that carbohydrate residues play in the function of these biopolymers has until recently remained obscure. In the last 15 years evidence has accumulated that supports the participation of the carbohydrate portion of these complexes in cell-cell recognition (3), cell adhesion (4) and cell transformation processes (5). In addition, an increasing body of evidence now suggests that an initial step in the assimilation of glycoproteins involves the binding of specific carbohydrate residues of glycoproteins, called recognition markers, to specific cell surface receptors.

The elegant pioneering experiments of Ashwell, Morell and co-workers (6) demonstrated that glycoproteins terminating in β-galactosyl residues are rapidly and selectively removed from the circulation by mammalian hepatocytes and that specific receptors for β-galactosyl residues occur on the plasma membranes (7). Other carbohydrate residues also serve as recognition markers for the cellular assimilation of specific glycoproteins. Lunney and Ashwell (8) have shown that avian hepatocytes specifically assimilate circulating glycoproteins that contain oligosaccharide chains terminating in β-N-acetylglucosaminyl residues. Mannosyl residues have been implicated as putative recognition markers for binding of gonadotropin to rat testis cells (9), and the clearance of ribonuclease b (10), agalacto-orosomucoid (11), and β-glucuronidase (12) from the circulation by rat liver.

In the late 1960's an elegant series of experiments by Neufeld and colleagues demonstrated that "corrective factors" obtained from normal skin fibroblasts or human urine eliminated

*This work was supported in part by Grant AM 10531 from The National Institute of Arthritis, Metabolic and Digestive Diseases, National Institutes of Health, and grants from the National Foundation-March of Dimes and from the Arthritis Foundation, Michigan Chapter.

[1] Unless otherwise noted, all sugars are of the D-configuration.

0-8412-0466-7/79/47-088-163$05.75/0

the abnormal accumulation of glycosaminoglycan deposits that occur
in fibroblasts derived from patients with mucopolysaccharidoses
(13). The corrective factors were subsequently identified as
lysosomal enzymes that were absent or defective in the mutant
cells under study. These findings prompted a number of laboratories to attempt enzyme replacement therapy for the correction of
lysosomal storage diseases (14). While the results of such clin-
ical studies have in large part been disappointing, they have
stimulated attempts to elucidate the mechanism by which lysosomal
enzymes, added externally to cells, find their way to the intracellular storage materials.

Several laboratories have studied the assimilation of specific lysosomal enzymes using as model systems skin fibroblasts
deficient in the enzyme under study. The underlying mechanism for
the translocation of lysosomal enzymes was hypothesized to involve
binding of carbohydrate-containing recognition markers to specific
cell surface receptors (15). In support of this hypothesis Hickman, Shapiro, and Neufeld (16) found that treatment of N-acetyl-
β-hexosaminidase with periodate under conditions that did not
affect enzymatic activity prevented the efficient assimilation of
this enzyme by Sandhoff fibroblasts. Additionally, Kresse and von
Figura (17) found that treatment of N-acetyl-α-hexosaminidase with
β-galactosidase reduced the assimilation of this enzyme by San-
filippo B fibroblasts.

Evidence for participation of carbohydrate residues in the
assimilation of β-galactosidase by generalized gangliosidosis
fibroblasts has recently been obtained in the authors' laboratories. Bovine testicular β-galactosidase, a lysosomal enzyme, was
highly purified by affinity chromatography and found to be a glycoprotein containing approximately 18 residues of mannose and 10
residues of N-acetylglucosamine per molecule (18,19,20). The
enzyme was rapidly assimilated by the β-galactosidase-deficient
generalized gangliosidosis fibroblasts. When β-galactosidase was
treated with a partially purified mannosidase preparation from
Aspergillus niger, mannose residues were removed from the enzyme
with concomitant loss of assimilability (21). These results
suggested that mannosyl residues either comprised or contributed
to the recognition marker of the enzyme. This observation was
strengthened by the observation that mannose and methyl α- or β-
mannosides inhibited the assimilation of β-galactosidase whereas
other monosaccharides or methyl glycosides did not (21). By
analogy to the well established binding of serum glycoproteins to
liver cell receptors, it was hypothesized that the assimilation of
β-galactosidase proceeded by the binding of mannosyl residues to
specific cell surface receptors (Figure 1). The above hypothesis
prompted experiments designed to explore the kinetic and structural parameters associated with the assimilation of β-galactosidase by generalized gangliosidosis fibroblasts.

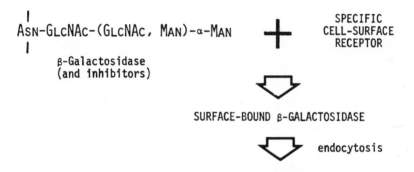

Figure 1. *Hypothetical mechanism for the selective assimilation of bovine testicular β-galactosidase by generalized gangliosidosis fibroblasts*

Characteristics of the Fibroblast/β-Galactosidase Assimilation System. Preliminary experiments demonstrated the assimilation of the enzyme was linear with time for periods of up to 12
hours at which time approximately 10% of the added enzyme was
associated with the cells. Furthermore, on exposure of the fibroblasts to β-galactosidase for 12 hours, the enzyme was demonstrated in the interior of the cells by histochemical staining (Figure
2). That the assimilated β-galactosidase was functional and
participated in intracellular catabolic metabolism was demonstrated by its correction of abnormal accumulation of sulfate-containing materials in enzyme deficient fibroblasts (23). When increasing amounts of enzyme were added to culture medium, the rate of
enzyme assimilation became constant at levels approximating 50
units/ml growth medium (Figure 3). Saturation of the fibroblasts
with β-galactosidase would be expected if assimilation were
mediated by adsorption of the enzyme to a limited number of cell
receptors (15,24). If assimilation of β-galactosidase were
mediated by passive fluid endocytosis, a process that does not
require participation of cell surface receptors, saturation of the
cells with enzyme should not occur. Horseradish peroxidase has
been used to measure the rate of passive fluid endocytosis (25).
As expected, the assimilation of horseradish peroxidase by generalized gangliosidosis fibroblasts was found to be a linear function of the concentration of the enzyme in the growth medium
(Figure 4).
      The rapid and selective assimilation of β-galactosidase by
generalized gangliosidosis fibroblasts provides additional evidence for the participation of cell surface receptors. Calcula-
tion of the percent internalization of β-galactosidase and
peroxidase at limiting concentrations (from the results presented
in Figures 3 and 4) revealed that β-galactosidase was assimilated
at about 450 times the rate of horseradish peroxidase. This
selective internalization of β-galactosidase may be due to the
presence of specific receptors for the enzyme. Thus, while cell
surface receptors for β-galactosidase have not yet been isolated
and identified, criteria for the participation of an initial
absorptive step during endocytosis have been satisfied.

      Characterization of the Carbohydrate Structures of β-Galacto-
sidase. In order to elucidate the chemical structure of the
recognition marker, glycopeptides obtained from β-galactosidase
and related glycoproteins were studied. Since sufficient quantities of highly purified β-galactosidase were not initially available, attention, therefore, was focused on an abundant glycoprotein fraction obtained as a by-product in the preparation of
β-galactosidase. The glycoprotein fraction, subsequently called
"inhibitor glycoprotein fraction," was a potent inhibitor of
β-galactosidase assimilation, presumably because of its competi-
tion with β-galactosidase for the receptor sites contained on the
cell surface. The inhibitor glycoprotein fraction passed through

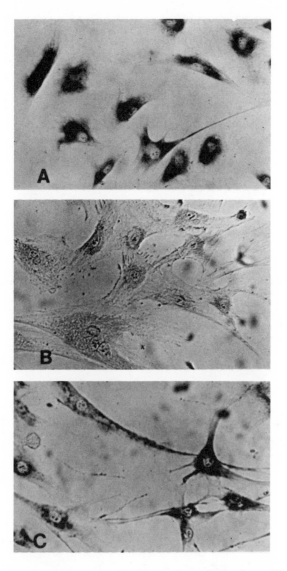

Figure 2. Histochemical stain of fibroblasts for β-galactosidase. The fibroblasts were plated on glass slides and stained for β-galactosidase with 5-bromo-4-chloro-3-indolyl β-galactoside reagent as described by Lake (22). A, normal fibroblasts; B, generalized gangliosidosis fibroblasts; and C, generalized gangliosidosis fibroblasts exposed to β-galactosidase (50 units/mL medium) for 12 hours prior to staining; the presence of major enzyme activity in a perinuclear region indicates the enzyme does not coat the cell surface but rather is internalized, presumably in in the cell lysosomes.

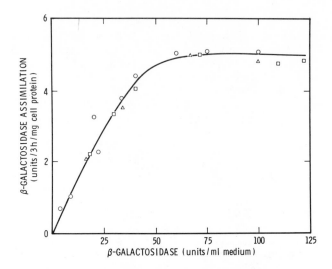

Figure 3. Effect of the concentration of β-galactosidase in the medium on the rate of assimilation of β-galactosidase by generalized gangliosidosis fibroblasts. The different symbols represent experiments performed at different times using the same enzyme preparation and fibroblast cell strain. The specific activity of the enzyme was approximately 4000 units/mg protein. One unit of enzyme is that amount which catalyzes the hydrolysis of 1 μmole p-nitrophenyl β-galactopyrano-side per minute at pH 4.3 and 37° (18).

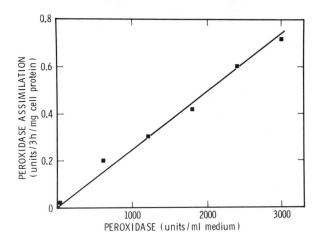

Figure 4. Effect of the concentration of horseradish peroxidase on the rate of assimilation of this enzyme by generalized gangliosidosis fibroblasts. The specific activity of the enzyme was approximately 3000 units/mg protein.

the affinity support used to purify the β-galactosidase and was further purified by adsorption to concanavalin A-Sepharose and elution with methyl α-mannoside (21). The resulting inhibitor glycoprotein fraction had a carbohydrate composition similar to that of β-galactosidase (Table I) and contained several other lysosomal enzymes that were present in the crude extract (Table II).

A glycopeptide fraction was prepared from the inhibitor glycoprotein fraction (and subsequently from β-galactosidase) by digestion with pronase. The resulting glycopeptide fraction was purified by gel filtration, adsorbed on concanavalin A-Sepharose and eluted with acetic acid. The glycopeptide fraction that bound to the concanavalin A-Sepharose column inhibited the assimilation of β-galactosidase by generalized gangliosidosis fibroblasts. This fraction was further purified by sequential gel filtration and by ion exchange chromatography procedures and analyzed. Preliminary studies showed the glycopeptide fraction did not undergo alkaline β-elimination indicating the probable attachment of the oligosaccharide chains to the polypeptide by N-glycosidic bonds between C-1 of acetylglucosamine and asparagine residues (29). The carbohydrate composition of the glycopeptide fraction is presented in Table III. Analysis of unit A glycopeptide fraction prepared from thyroglobulin by the procedure of Arima, Spiro, and Spiro (30) is also presented. As may be seen the thyroglobulin derivative has a carbohydrate composition similar to that found for the inhibitor glycopeptide fraction.

The results of controlled enzymatic degradation of the glycopeptides derived from the inhibitor glycoprotein fraction are summarized in Figure 5; the glycopeptide fraction obtained from β-galactosidase, and the unit A glycopeptide fraction from thyroglobulin gave similar results. Taken together these studies suggest each glycopeptide fraction contained terminal α-mannosyl residues. At least a portion of the oligosaccharide chains terminated in α-1,2-linked mannosyl residues. Mannose was separated from the peptide portion by N-acetylchitobiosyl residues; a single β-mannosyl residue was attached to the N-acetylchitobiose. No organic phosphate was observed in the glycopeptides by colorimetric analysis; the sensitivity of the methodology was such that less than one phosphate residue per 10 residues of aspartic acid in the glycopeptides would have been detected.

Inhibition of Assimilation of β-Galactosidase by Mannose-Rich Compounds. The results of inhibition studies utilizing the glycopeptides, glycoproteins and other mannose derivatives are summarized in Figure 6. The percent inhibition of β-galactosidase assimilation is plotted as a logarithmic function of the mannose concentration of the compound added to the growth medium. As may be seen glycopeptides derived from the inhibitor glycoprotein fraction were approximately 100 times more inhibitory than synthetic methyl α- or β-mannosides. These results suggest that at

Table I

CARBOHYDRATE ANALYSIS OF BOVINE TESTICULAR
β-GALACTOSIDASE AND INHIBITOR GLYCOPROTEINS*

	β-Galactosidase	Inhibitor glycoproteins
	μmoles/mg protein	μmoles/mg protein
Mannose	0.27	0.32
N-Acetylglucosamine	0.15	0.17
Galactose	<0.01	0.06
N-Acetylneuraminic acid	<0.01	0.05
L-Fucose	<0.01	0.01

*Neutral sugars and hexosamines were estimated after Dowex-50 (H$^+$)-catalyzed hydrolysis by gas liquid chromatography of glycitol acetate derivatives (26,27). N-Acetylneuraminic acid was measured by a periodate-resorcinol procedure (28).

Table II

ANALYSIS OF INHIBITOR GLYCOPROTEINS FROM BOVINE TESTES EXTRACT

Analyses	Crude supernatant	Con-A Sepharose eluant	Recovery
	total units*	total units*	%
Lysosomal enzymes			
β-Mannosidase	22	16	74
Arylsulfatase A	135	44	33
β-N-Acetylhexosaminidase	3,070	825	27
β-Glucuronidase	72	17	23
Hyaluronidase	239,000	43,200	18
α-Mannosidase	44	6	14
Acid phosphatase	81	7	9
	mg	mg	%
Protein	15,780	96	0.6

*Units = μmoles/min except hyaluronidase = National Formulary units

Table III

ANALYSIS OF GLYCOPEPTIDE PREPARATIONS

Analyses	Inhibitor glycopeptides	β-Galactosidase glycopeptides	Thyroglobulin (unit A)
	moles*	moles*	moles*
Aspartic acid	1.00	1.00	1.00
Glucosamine	1.93	2.22	2.10
Mannose	4.62	4.15	5.51
Mannose released by jack bean α-mannosidase	3.47	3.06	4.50

*Relative to aspartic acid (amino acid analyzer). Mannose and glucosamine were estimated after Dowex 50 (H^+)-catalyzed hydrolysis by gas liquid chromatography of glycitol acetate derivatives (26,27); enzymatically liberated mannose was estimated by omission of the hydrolysis step.

least a portion of the recognition marker is associated with the carbohydrate-containing area of the parent glycoproteins. It should be emphasized, however, that the observed inhibition with the inhibitor glycopeptide fraction is less than 10% of that observed with the parent inhibitor glycoproteins. These results suggest that the recognition marker of the glycopeptide is somewhat altered or that its action is modified by the associated polypeptide chain of the parent glycoproteins. In this regard, Bahl (9) has suggested that the polypeptide portion of gonadotropin, a mannose-containing glycoprotein, plays a role in its binding to testes membrane preparations. The observed inhibition of β-galactosidase assimilation by the mannose-rich thyroglobulin unit A glycopeptide fraction was not unexpected in view of its similarity in terms of enzymatic and chemical analysis to the inhibitor glycopeptide fraction. However, unit B glycopeptide fraction derived from thyroglobulin, which contains substantial amounts of galactose and sialic acid, did not inhibit β-galactosidase assimilation at similar mannose concentrations. Synthetic α-D-mannopyranosyl-(1→2)-D-mannose gave inhibition values intermediate between those observed with methyl mannosides and the inhibitor glycopeptide fraction. On the basis of these results it is tempting to speculate that the α-(1→2)-linked mannosyl residues, found in the inhibitor glycoprotein fraction, comprise a characteristic portion of the recognition marker of lysosomal enzymes; this linkage has apparently not been reported in glycoprotein carbohydrate chains

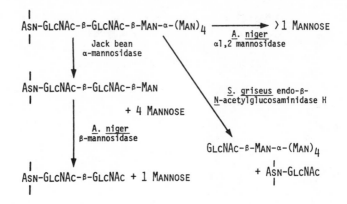

Figure 5. *Enzymatic analysis of inhibitor glycopeptides. Glycopeptide and oligosaccharide products were isolated by ion-exchange and gel filtration techniques. Monosaccharide constituents were determined by GLC of their glycitol acetate derivatives following Dowex-50 [H⁺] catalyzed acid hydrolysis (26, 27). Liberated mannose was determined by direct GLC of its glycitol acetate; A. niger α-1,2-mannosidase and β-mannosidase were prepared by the procedures of Swaminathan et al. (31) and Elbein et al. (32), respectively; S. griseus endoglucosaminidase was prepared as described by Tarentino and Maley (33).*

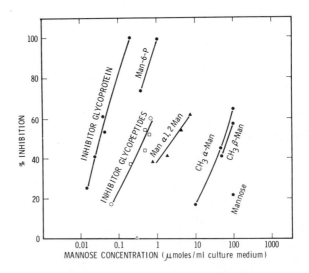

Figure 6. *Inhibition of β-galactosidase assimilation by inhibitor glycoproteins, glycopeptides, and derivatives. The open symbols represent data obtained with glycopeptides; open squares, inhibitor glycoprotein fraction; open circles, unit A glycopeptides prepared from thyroglobulin (30); and open triangle, unit B glycopeptides from thyroglobulin (30).*

terminating in galactose and sialic acid residues.
   In a recent report, Kaplan, Achord, and Sly (34) while studying the assimilation of human platelet β-glucuronidase by fibroblasts made the surprising observation that mannose-6-phosphate
inhibited the assimilation of β-glucuronidase at concentrations
several orders of magnitude lower than that observed with free
mannose. Furthermore, treatment of the enzyme with alkaline
phosphatase destroyed its ability to be assimilated by fibroblasts.
These results prompted the investigators to suggest that phosphate
residues were an integral part of the recognition marker of this
enzyme. Similar results have been obtained in studies of the
assimilation of urinary α-iduronidase (15). As shown in Figure 6
mannose-6-phosphate also strongly inhibits the assimilation of
bovine testicular β-galactosidase. In addition, treatment of
β-galactosidase with alkaline phosphatase was found to destroy the
assimilation of this enzyme by fibroblasts (data not shown). These
observations indicate phosphate residues may also be involved in
the assimilation of β-galactosidase.

   Attempts to Detect Mannose-6-Phosphate in β-Galactosidase and
Inhibitor Glycopeptides. Although colorimetric analysis of the
inhibitor glycopeptide fraction for phosphate proved negative, the
occurrence of trace quantities of mannose phosphate could not be
eliminated. Experiments were therefore undertaken to demonstrate
the possible presence of small amounts of mannose-6-phosphate in
β-galactosidase and in the inhibitor glycopeptide fraction.
   Advantage was taken of the recognized acid stability of
mannose-6-phosphate (35). Glycopeptides or glycoproteins that
were examined for mannose-6-phosphate were subjected to Dowex-50
(H^+)-catalyzed acid hydrolysis (26) under conditions that released
maximal amounts of free mannose but hydrolyzed less than 50% of
added mannose-6-phosphate. The resulting free sugars and sugar
phosphates were reduced with NaB^3H_4 (36) and subjected to paper
electrophoresis. As shown in Figure 7, mannose-6-phosphate
treated in this manner yielded approximately equal quantities of
[^{3}H]mannitol and [^{3}H]mannitol-6-phosphate. Furthermore, as shown
in Figure 7 [^{3}H]mannitol-6-phosphate was readily demonstrated
after hydrolysis and reduction of a yeast phosphomannan that
contains bound mannose-6-phosphate (37). When glycopeptides
derived from β-galactosidase or the inhibitor glycoprotein fraction were subjected to this treatment only small amounts of ^{3}H
were found to comigrate with authentic mannitol-6-phosphate (peak
C, Figure 8). On elution and re-electrophoresis of peak C in
borate buffer no ^{3}H was found to comigrate with authentic mannitol
-6-phosphate. The sensitivity of these experiments was such that
one mannose-6-phosphate residue per approximately 750 mannose
residues of the glycopeptide fraction would have been detected.
Other experiments were carried out with intact β-galactosidase
(Figure 9). Similar results were obtained; in this case, less
than one mannose-6-phosphate residue would have been detected for

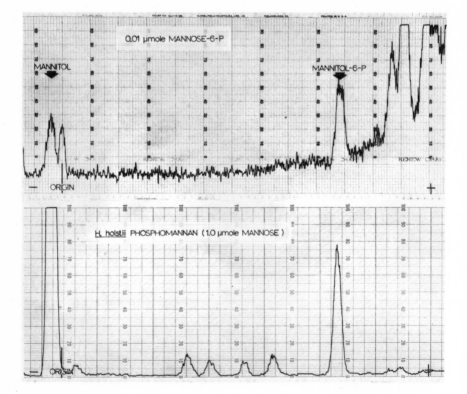

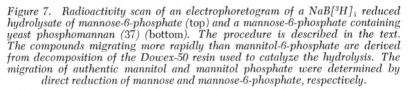

Figure 7. Radioactivity scan of an electrophoretogram of a NaB[³H]₄ reduced hydrolysate of mannose-6-phosphate (top) and a mannose-6-phosphate containing yeast phosphomannan (37) (bottom). The procedure is described in the text. The compounds migrating more rapidly than mannitol-6-phosphate are derived from decomposition of the Dowex-50 resin used to catalyze the hydrolysis. The migration of authentic mannitol and mannitol phosphate were determined by direct reduction of mannose and mannose-6-phosphate, respectively.

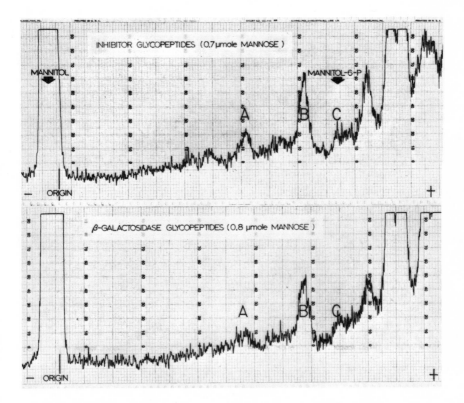

Figure 8. Radioactivity scan of a paper electrophoretogram of NaB[³H]₄ reduced hydrolysate of inhibitor glycopeptides (top) and glycopeptides from β-galactosidase (bottom). Phosphorylated compounds could not be detected in peaks A and B or in the mannitol-6-phosphate region, C. (see the text for details).

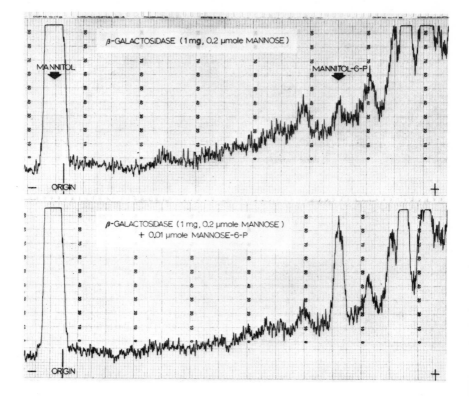

Figure 9. *Radioactivity scan of a paper electrophoretogram of NaB[^{3}H]$_4$ reduced hydrolysate of β-galactosidase (top) and a mixture of β-galactosidase and mannose-6-phosphate (bottom). Experimental details are given in the text.*

every 250 mannose residues present in the β-galactosidase preparation. Addition of mannose-6-phosphate to the β-galactosidase
before hydrolysis gave the expected amount of $[^3H]$mannitol-6-phos-
phate indicating the sugar phosphate was stable to the hydrolysis
and reduction conditions employed (Figure 9).
     Acidic ^3H-labeled compounds (A and B, Figure 8) were observed
on the paper electrophoretograms after hydrolysis and reduction of
the inhibitor glycopeptide fraction. The ^3H-labeled compounds
were not detected in control samples but were observed in lower
amounts in hydrolysates derived from β-galactosidase (Figure 9).
Strong mineral acid hydrolysis of A and B released a ^3H-labeled
compound tentatively identified as ^3H-mannitol by paper chromatography and paper electrophoresis. The electrophoretic migration
of A and B was not altered by treatment of the compounds with
alkaline phosphatase under conditions in which ^3H-mannitol-6-phos-
phate was completely cleaved; it therefore seems unlikely that A
and B are phosphorylated compounds.

Discussion

     The results of kinetic studies have demonstrated that the
assimilation of β-galactosidase by generalized gangliosidosis
fibroblasts conforms to the criteria established for a receptor-
mediated adsorptive endocytosis (15,24). Two lines of evidence
point to a role for mannose in the proposed recognition marker on
the enzyme. Enzymatic removal of mannose residues from the enzyme
is accompanied by a loss in assimilability of the enzyme, and
enzyme assimilation is inhibited by a variety of compounds containing terminal mannosyl residues; these include methyl glycosides, oligosaccharides, glycoproteins, and glycopeptides.
     Structural studies of the carbohydrate residues of β-galacto-
sidase and related inhibitor glycoproteins indicate that the oligosaccharide chains contain glycosidically-bound mannose and N-
acetylglucosamine residues linked in the manner shown below:

$$Asn \overset{|}{\underset{|}{\quad}} GlcNAc \overset{\beta}{\longleftarrow} GlcNAc \overset{\beta}{\longleftarrow} Man \overset{\alpha}{\longleftarrow} (Man)_n$$

Similar mannose-rich oligosaccharide residues have been described
in thyroglobulin (30), in cell surface glycoproteins (38), and in
immunoglobulins (39). Galactose, fucose and sialic acid were not
detected in the purified β-galactosidase suggesting that these
sugars do not contribute to the recognition marker.
     Evidence has been obtained that supports the suggestion of
Kaplan, Achord, and Sly (35) that phosphate residues contribute to
the assimilation of lysosomal enzymes by fibroblasts. The strong
inhibition of β-galactosidase assimilation by mannose-6-phosphate
may not be explained solely on the basis of the compound's mannose
content. Experiments in our laboratories designed to detect the
presence of mannose-6-phosphate in β-galactosidase have demon-

strated less than one residue of mannose-6-phosphate for every 250 residues of mannose. Since each enzyme molecule contains an average 18 residues of mannose, less than 10 percent of the enzyme molecules could contain a single mannose-6-phosphate residue. However, these experiments were not designed to detect phosphate substituted at positions other than at the C-6 of mannose. Davis et al. (40,41) have presented evidence for the presence of phosphate substituted on other than the C-6 position of mannose in a rat brain glycoprotein fraction. Alternatively, it seems possible that other constituents of the β-galactosidase such as the polypeptide chain may be phosphorylated.

The observation that phosphate-free glycopeptides at low concentrations effectively inhibit the assimilation of β-galactosidase by fibroblasts implies that details of the carbohydrate structure play a role in enzyme recognition by cells. The role of phosphate in the assimilation of lysosomal enzymes remains obscure. Conceivably, phosphate may lend additional specificity to the assimilation system, or perhaps it could play a role in events subsequent to the initial binding of the enzyme to the cell surface.

It seems plausible, in the light of the known complexity of mammalian glycoproteins, that additional specific recognition systems will be found. The elucidation of the chemical fine structure of recognition markers and the coupling of specific recognition markers to proteins may allow the directed cellular assimilation of specific proteins. The successful achievement of these goals will 1) enhance our understanding of the process of adsorptive endocytosis, 2) may elucidate mechanisms for the regulation of turnover and distribution of glycoproteins, lysosomal enzymes, and hormones and 3) may ultimately lead to effective enzyme replacement therapy for treatment of the many devastating inherited lysosomal storage disorders.

Literature Cited

1. Gottschalk, A., in "Glycoproteins" (Gottschalk, A., ed.) pp. 1-23, Elsevier, New York (1972).
2. Gottschalk, A., in "Glycoproteins" (Gottschalk, A., ed.) pp. 24-30, Elsevier, New York (1972).
3. Walther, B.T., Öhman, R. and Roseman, S., Proc. Nat. Acad. Sci. U.S.A. (1973) 70, 1569-1573.
4. Chipowsky, S., Lee, Y.C. and Roseman, S., Proc. Nat. Acad. Sci. U.S.A. (1973) 70, 2309-2312.
5. Buck, C.A., Fuhrer, J.P., Soslau, G. and Warren, L., J. Biol. Chem. (1974) 249, 1541-1550.
6. Ashwell, G. and Morell, A.G., Adv. Enzymol. (1974) 41, 99-128.
7. Hudgin, R.L., Pricer, W.E., Jr., Ashwell, G., Stockert, R.J. and Morell, A.G., J. Biol. Chem. (1974) 249, 5536-5543.

8. Lunney, J. and Ashwell, G., Proc. Nat. Acad. Sci. U.S.A. (1976) 73, 341-343.
9. Bahl, O.P., Fed. Proc. (1977) 36, 2119-2127.
10. Baynes, J.W. and Wold, F., J. Biol. Chem. (1976) 251, 6016-6024.
11. Achord, D.T., Brot, F.E. and Sly, W.S., Biochem. Biophys. Res. Commun. (1977) 77, 409-415.
12. Sly, W.S., J. Supramolecular Structure (1977) 6 (Supp. 1),36.
13. Neufeld, E.F., Lim, T.W. and Shapiro, L.J., Ann, Rev. Biochem. (1975) 44, 357-376.
14. "Enzyme Therapy in Genetic Diseases," Birth Defects: Original Article Series, Vol. IX (Bergsma, D., ed.) The Williams and Wilkins Co., Baltimore (1973).
15. Neufeld, E.F., Sando, G.N., Garvin, A.J. and Rome, L.H., J. Supramolecular Structure (1977) 6, 95-101.
16. Hickman, S., Shapiro, L.J. and Neufeld, E.F., Biochem. Biophys. Res. Commun. (1974) 57, 55-61.
17. Kresse, H. and von Figura, K., in "Enzyme Therapy in Lysosomal Storage Diseases" (Tager, J.M., Hooghwinkel, G.J.M. and Daems, W. T., eds.) pp. 173-174, Elsevier, New York (1974).
18. Distler, J.J. and Jourdian, G.W., J. Biol. Chem. (1973) 248, 6772-6780.
19. Distler, J.J. and Jourdian, G.W., Arch. Biochem. Biophys. (1977) 178, 631-643.
20. Distler, J.J. and Jourdian, G.W., Methods Enzymol. (1978) 50, 514-520.
21. Hieber, V., Distler, J., Myerowitz, R., Schmickel, R.D. and Jourdian, G.W., Biochem. Biophys. Res. Commun. (1976) 73, 710-717.
22. Lake, B.D., Histochem. J. (1974) 6, 211-218.
23. Distler, J., Hieber, V., Schmickel, R., Myerowitz, R. and Jourdian, G.W., in "Disorders of Connective Tissue," Birth Defects: Original Article Series, Vol. XI (Bergsma, D., ed.) pp. 311-315, Symposia Specialists, Miami (1975).
24. Jacques, P.J., in "Lysosomes in Biology and Pathology," Vol. II (Dingle, J.T. and Fell, H., eds.) pp. 395-420, North Holland, Amsterdam (1969).
25. Steinman, R.M., Silver, J.M. and Zanvil, A.C., J. Cell Biol. (1974) 63, 949-969.
26. Lehnhardt, W.F. and Winsler, R.J., J. Chromatogr. (1968) 34, 471-479.
27. Porter, W.H., Anal. Biochem. (1975) 63, 27-43.
28. Jourdian, G.W., Dean, L. and Roseman, S., J. Biol. Chem. (1971) 246, 430-435.
29. Marshall, R.D. and Neuberger, A., Biochemistry (1964) 3, 1596-1600.
30. Arima, T., Spiro, M.J. and Spiro, R.G., J. Biol. Chem. (1972) 247, 1825-1835.
31. Swaminathan, N., Matta, K.L., Donoso, L.A. and Bahl, O.P., J. Biol. Chem. (1972) 247, 1775-1779.

32. Elbein, A.D., Adya, S. and Lee, Y.C., J. Biol. Chem. (1977)
    252, 2026-2031.
33. Tarentino, A.L. and Maley, F., J. Biol. Chem. (1974) 249,
    811-817.
34. Kaplan, A., Achord, D.T., and Sly, W.S., Proc. Nat. Acad.
    Sci. U.S.A. (1977) 74, 2026-2030.
35. Leloir, L.F. and Cardini, C.E., Methods Enzymol. (1957) 3,
    840-850.
36. Thieme, T.R. and Ballou, C.E., Biochemistry (1972) 11, 4121-
    4129.
37. Slodki, M.E., Biochim. Biophys. Acta (1962) 57, 525-533.
38. Muramatsu, T., Koide, N., Ceccarini, C. and Atkinson, P.H.,
    J. Biol. Chem. (1976) 251, 4673-4679.
39. Hickman, S., Kornfeld, R., Osterland, C.K. and Kornfeld, S.,
    J. Biol. Chem. (1972) 247, 2156-2163.
40. Davis, L.G., Javaid, J.I. and Brunngraber, E.G., F.E.B.S.
    (Fed. Eur. Biochem. Soc.) Lett. (1976) 65, 30-34.
41. Davis, L.G., Costello, A.J.R., Javaid, J.I. and Brunngraber,
    E.G., F.E.B.S. (Fed. Eur. Biochem. Soc.) Lett. (1976) 65,
    35-38.

Note added in proof: Since submission of the original manuscript,
a small amount of mannose-6-phosphate (1-2% of the total mannose)
has been detected in glycopeptides obtained from a trypsin digest
of the inhibitor glycoprotein fraction. At equivalent mannose
concentrations the tryptic glycopeptides inhibit β-galactosidase
assimilation about 5 times greater than the pronase glycopeptide
fraction. These results may indicate that pronase alters the
recognition marker of the glycopeptides.

    This work was supported in part by Grant AM 10531 from The
National Institute of Arthritis, Metabolic and Digestive Diseases,
National Institutes of Health, and grants from the National Found-
ation-March of Dimes and from the Arthritis Foundation, Michigan
Chapter.

RECEIVED September 8, 1978.

The Absence of Carbohydrate Specific Hepatic Receptors for Serum Glycoproteins in Fish

GILBERT ASHWELL

National Institute of Arthritis, Metabolism, and Digestive Diseases, National Institutes of Health, Bethesda, MD 20014

RAYMOND P. MORGAN II[1]

The Chesapeake Biological Laboratory, University of Maryland Center for Environmental and Estuarine Studies, Solomons, MD 20688

In mammalian and avian species, specific carbohydrate moieties of serum glycoproteins have been shown to be of critical importance for their continued survival in the circulation. The removal of terminal sialic acid from these serum proteins exposes the penultimate, non-reducing sugar, galactose. The resulting asialoglycoprotein, upon intravenous injection into mammals, is rapidly cleared from the circulation and catabolized in the liver (1). The hepatic receptor responsible for binding and uptake has been purified to homogeneity and shown to bind specifically to the exposed galactose residues (2, 3). In contrast, birds and reptiles which are deficient in this receptor possess an alternate hepatic binding protein specific for terminal N-acetylglucosamine residues. The latter binding protein has also been isolated in pure form and shown to bind specifically to those glycoproteins from which both sialic acid and galactose have been removed to expose the underlying N-acetylglucosamine residues (agalactoglycoproteins) (4, 5).

In view of the presumed role of these unique proteins in the regulation of serum glycoprotein homeostasis, it became of interest to determine whether fish, a more primitive evolutionary species, exhibited a similar or divergent control mechanism.

Materials and Methods

Human α_1-acid glycoprotein (orosomucoid) was provided by Dr. M. Wickerhauser of the American Red Cross Research Center, Bethesda, Md.; bovine serum albumin was purchased from Armour Pharmaceutical Co.

Clostridium perfringens neuraminidase was purchased from Worthington Co. and purified by affinity chromatography (6). β-Galactosidase and β-N-Acetylglucosaminidase were isolated from a culture filtrate of Diplococcus pneumoniae by a modification of the procedure of Hughes and Jeanloz (7).

Sequential removal of the terminal sugars from orosomucoid with neuraminidase, β-galactosidase, and β-N-acetylglucosaminidase

[1] Current address: Battelle Columbus Laboratories, William F. Clapp Laboratories, Inc., Duxbury, MA 02332.

resulted in the formation of asialo-, agalacto-, and ahexosamino-
derivatives, respectively. After each step the modified glycoproteins were isolated by Sephadex G-150 chromatography and the
release of the individual monosaccharides was monitored prior to
further enzymatic degradation as described previously (5).
$Na^{125}I$, carrier free, in 0.1N NaOH was obtained from New England
Nuclear Co. Orosomucoid, or its appropriate derivative (100 µg)
was iodinated with 1 mC of $Na^{125}I$ by a modificaiton of the method
of Greenwood et al. (8). The iodinated glycoproteins, purified
by passage through a column of Sephadex G-25, were recovered
with specific activities ranging from 0.5 to 0.8 µCi/µg.

Striped bass (morone soxatilis, Percichthyidae) were collected from the Patuxtent River, Maryland, with an 8 m otter trawl.
After collection, the fish were immediately transported back to
the Chesapeake Biological Laboratory. The fish were held in a
flowing water system with ambient temperature, salinity and dissolved oxygen. Obviously damaged specimens were discarded as
well as any with fungal infections. The fish, held for approximately one week prior to experimentation, were older than 1 year
but less than 3 years. Weights ranged from 400-800 grams.

The striped bass were injected (26 gauge 3/8" needle, 1 ml
syringe) with 0.1 ml of the several glycoprotein derivatives
and/or bovine serum albumin through a gill vein. The amounts
injected ranged from 1-5 µg of protein containing from 1-5 X
10^6 cpm. Following injection, blood was withdrawn from the heart
with a 22 gauge 1-1/2" needle at varying time intervals; 0.5 to
1.0 ml of whole blood was collected per bleeding, transferred to
clean vials and allowed to clot. Approximately, 25 minutes after
the initial injection, the fish were killed and samples of the
liver, kidney and spleen were removed for radioactivity determination. Aliquots of the serum, recovered by centrifugation of the
clotted blood, were monitored for radioactivity on a Packard
Autogamma Spectrometer (60% efficiency).

Results and Discussion

The rapid, carbohydrate dependent clearance of specifically
modified serum glycoproteins, observed in mammalian species (1),
was not demonstrable in fish. Of the five proteins examined in
Fig. I, no significantly increased rate of disappearance from
the circulation could be correlated with the nature of the terminal, non-reducing glycoside moiety. The marginally faster clearance observed in the case of asialo-orosomucoid, although repro-
ducible, could not be ascribed to the participation of a specific
hepatic receptor. This conclusion is supported by the data in
Table I whereby it is evident that there was no selective accumulation of radioactivity in the liver; the major portion of counts
recoverable 25 minutes after injection were located in the
kidneys.

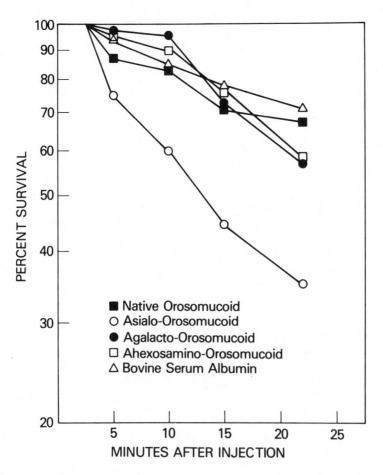

Figure 1. *Time course of clearance from the circulation of variously modified proteins. The points represent the average values obtained from two to five separate experiments.*

Table I. Tissue Distribution of Radioactivity at 25 min After Injection

Tissue	Protein Injected				
	Orosomucoid	Bovine Serum Albumin	Asialo-orosomucoid	Agalacto-orosomucoid	Ahexosamino-orosomucoid
	cpm/gm wet weight X 10^{-3}				
Liver	35	27	80	230	188
Kidney	185	158	124	1,675	1,524
Spleen	33	7	42	62	60

Further confirmation of the absence of a receptor analogous to that of mammalian and avian species was obtained from binding studies carried out in vitro. Detergent solubilized preparations of freshly prepared homogenates, or acetone dried powders, of fish liver and kidney were assayed for specific binding activity as described previously (2, 3). In no case was it possible to show selective binding of ^{125}I-asialo-orosomucoid or ^{125}I-agalacto-orosomucoid in excess of background values.

From the above evidence, it seems reasonable to infer that the hepatic receptors for galactose- and N-acetylglucosamine-terminated serum proteins present in mammals and birds, respectively, emerged at an evolutionary stage later than that of fish. The nature of the control mechanisms regulating the metabolism of serum glycoproteins in fish remains obscure.

Literature Cited

1. Ashwell, G. and Morell, A. G. (1974) Adv. Enzymol. 41, 99–128.
2. Hudgin, R. L., Pricer, W. E., Ashwell, G., Stockert, R. J., and Morell, A. G. (1974) J. Biol. Chem. 249, 5536–5543.
3. Kawasaki, T. and Ashwell, G. (1976) J. Biol. Chem. 251, 1296–1302.
4. Lunney, J. and Ashwell, G. (1976) Proc. Nat. Acad. Sci., USA 73, 341–343.
5. Kawasaki, T. and Ashwell, G. (1977) J. Biol. Chem. 252, 6536–6543.
6. Cuatrecasas, P. and Illiano, G. (1971) Biochem. Biophys. Res. Commun. 44, 178–184.
7. Hughes, R. C. and Jeanloz, R. W. (1964) Biochemistry 3, 1535–1543.
8. Greenwood, F. C., Hunter, W. M., and Glover, J. S. (1963) Biochem. J. 89, 114–123.

RECEIVED September 8, 1978.

14

Carbohydrate–Protein Interactions in Proteoglycans

L. ROSENBERG, H. CHOI, S. PAL, and L. TANG

Montefiore Hospital and Medical Center, Orthopedic and Connective Tissue
Research, 111 East 210th Street, Bronx, NY 10467

Proteoglycans are major structural components of
the intercellular matrix of connective tissues. In the
formation of proteoglycans, a proteoglycan basic unit
is first formed, called proteoglycan monomer. Proteoglycan monomer consists of glycosaminoglycan chains
covalently bound to a protein core. Several classes of
proteoglycans have been isolated from different connective tissues. These different classes of proteoglycans
are defined in terms of the kinds of glycosaminoglycan
chains which are bound to the protein core. In cartilages, proteoglycan monomer consists of chondroitin
sulfate and keratan sulfate bound to the same protein
core. In the intercellular matrix of blood vessel wall,
proteoglycan monomer consists of dermatan sulfate and
chondroitin sulfate bound to the same protein core. In
the plasma membranes of some cells, heparan sulfate is
bound to the protein core.
    In the intercellular matrix of cartilage, and per-
haps other tissues, most of the proteoglycan exists in
the form of proteoglycan aggregates, formed by the non-
covalent association of proteoglycan monomers with hyaluronic acid and link proteins. Carbohydrate-protein
interactions are involved in the binding of proteoglycan monomer to hyaluronate, and in the binding of link
protein to hyaluronate. Carbohydrate-protein interactions are also involved in the non-covalent association
of proteoglycans with collagen fibers in intercellular
matrix. The purpose of this review is to describe the
structure of proteoglycans, and to summarize the results
of recent studies of carbohydrate-protein interactions
between proteoglycan monomer and hyaluronate, link
protein and hyaluronate, and between proteoglycans and
collagen.

0-8412-0466-7/79/47-088-186$07.50/0
© 1979 American Chemical Society

Cartilage Proteoglycans

Cartilages are highly specialized connective tissues composed of relatively few cells distributed
throughout an abundant intercellular matrix. The intercellular matrix gives cartilage unusual mechanical properties essential for the normal function of diarthro-
dial joints. For example, because of the properties of
the intercellular matrix, articular cartilage is a relatively hard, yet elastic tissue. Articular cartilage
provides a smooth covering for the bony elements of di-
arthrodial joints and contributes to the almost fric-
tionless gliding of opposing joint surfaces. The intercellular matrix is composed mainly of collagen, proteoglycans and water. Collagen is an insoluble fibrous
protein with tensile strength. Proteoglycans are elastic molecules which tend to expand in solution and re-
sist compression into a smaller volume of solution.
The remarkable mechanical properties of articular cartilage result from the structure of collagen and proteoglycans, and from the properties of the fibrous compo-
site formed by the interactions of collagen and proteoglycans in intercellular matrix.
     A diagrammatic model of cartilage proteoglyan
monomer is shown in Figure 1. Cartilage proteoglycan
monomer consists of chondroitin sulfate and keratan
sulfate chains covalently bound to serine and threonine
residues within the protein core. Chondroitin sulfate
and keratan sulfate are members of the group of polysaccharides termed glycosaminoglycans. Glycosaminoglycans are composed of two different sugar residues which
alternate regularly in the polysaccharide chain. One
sugar residue is usually N-acetylgalactosamine or N-
acetylglucosamine. The other sugar residue is usually
glucuronic acid or iduronic acid. Thus glycosaminoglycans are composed of disaccharide repeating units. The
structures of the disaccharide repeating units of the
glycosaminoglycans, and of the linkage region of the
glycosaminoglycan chains to proteoglycan monomer core
protein, are shown in Figure 2. Chondroitin sulfate is
covalently bound to serine residues via the neutral
sugar trisaccharide, galactose-galactose-xylose (1-5).
Keratan sulfate is covalently bound mainly to threonine
and serine residues via N-acetylgalactosamine, to which
a sialylgalactosyl disaccharide is also attached (6-10).
     Proteoglycan monomers from different cartilages
vary in molecular weight and chemical composition, particularly in the relative amounts of chondroitin sulfate
and keratan sulfate. Indeed, proteoglycan monomers from
the same tissue are polydisperse and vary in size and

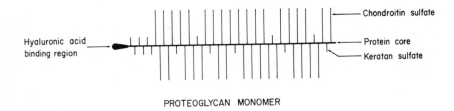

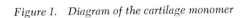

PROTEOGLYCAN MONOMER

Figure 1.   Diagram of the cartilage monomer

STRUCTURES OF THE GLYCOSAMINOGLYCANS AND THEIR LINKAGE REGIONS TO PROTEIN

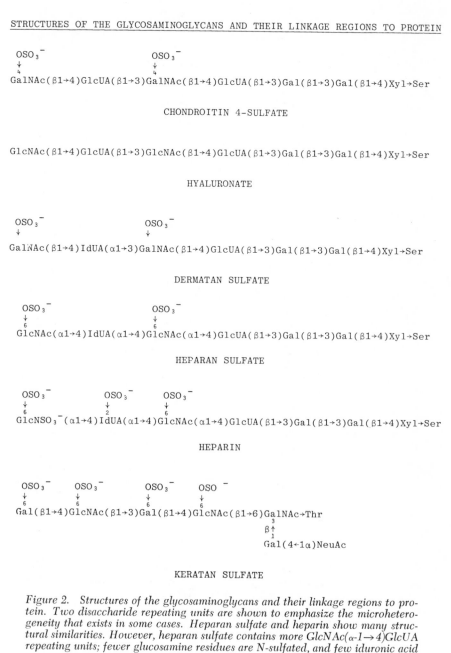

OSO_3^- OSO_3^-
↓   ↓
4   4
GalNAc(β1→4)GlcUA(β1→3)GalNAc(β1→4)GlcUA(β1→3)Gal(β1→3)Gal(β1→4)Xyl→Ser

CHONDROITIN 4-SULFATE

GlcNAc(β1→4)GlcUA(β1→3)GlcNAc(β1→4)GlcUA(β1→3)Gal(β1→3)Gal(β1→4)Xyl→Ser

HYALURONATE

OSO_3^- OSO_3^-
↓   ↓
GalNAc(β1→4)IdUA(α1→3)GalNAc(β1→4)GlcUA(β1→3)Gal(β1→3)Gal(β1→4)Xyl→Ser

DERMATAN SULFATE

OSO_3^- OSO_3^-
↓   ↓
6   6
GlcNAc(α1→4)IdUA(α1→4)GlcNAc(α1→4)GlcUA(β1→3)Gal(β1→3)Gal(β1→4)Xyl→Ser

HEPARAN SULFATE

OSO_3^- OSO_3^- OSO_3^-
↓   ↓   ↓
6   2   6
GlcNSO₃⁻(α1→4)IdUA(α1→4)GlcNAc(α1→4)GlcUA(β1→3)Gal(β1→3)Gal(β1→4)Xyl→Ser

HEPARIN

OSO_3^- OSO_3^- OSO_3^- OSO^-
↓   ↓   ↓   ↓
6   6   6   6
Gal(β1→4)GlcNAc(β1→3)Gal(β1→4)GlcNAc(β1→6)GalNAc→Thr
                                           3
                                           β↑
                                           1
                                     Gal(4←1α)NeuAc

KERATAN SULFATE

Figure 2. Structures of the glycosaminoglycans and their linkage regions to protein. Two disaccharide repeating units are shown to emphasize the microheterogeneity that exists in some cases. Heparan sulfate and heparin show many structural similarities. However, heparan sulfate contains more GlcNAc(α-1→4)GlcUA repeating units; fewer glucosamine residues are N-sulfated, and few iduronic acid residues are sulfated at C2.

composition. However, a representative proteoglycan
monomer from bovine articular cartilage would have a
protein core approximately 200,000 in molecular weight,
measuring 3000 Å long. To this would be attached about
100 chondroitin sulfate chains varying from 2×10^4 to
3×10^4 in molecular weight. Keratan sulfate chains
3×10^3 to 7×10^3 in molecular weight would also be
attached to the protein core. The entire proteoglycan
monomer would be approximately 2×10^6 to 3×10^6 in
molecular weight.

    <u>Structure of Proteoglycan Aggregates.</u> In the in-
tercellular matrix of cartilage, most of the proteoglycan exists in the form of aggregates of high molecular
weight. The molecular architecture of the cartilage
proteoglycan aggregate is shown in Figure 3. Hyaluronic
acid forms the filamentous backbone of the aggregate
(<u>11-19</u>). The aggregate is formed by the non-covalent
association of many proteoglycan monomers with hyaluronate. As indicated in the model of the proteoglycan
aggregate shown in Figure 3, proteoglycan monomer core
protein consists of three major regions which differ in
structure and function. One end of proteoglycan monomer
where proteoglycan monomer binds to hyaluronate, contains little or no chondroitin sulfate or keratan sulfate. It consists of a polypeptide about 60,000 in
molecular weight with a globular conformation. It contains the binding site of proteoglycan monomer core protein for hyaluronic acid. This region is called the
hyaluronic acid binding region of proteoglycan monomer
core protein.
    Most of the length of core protein, which extends
towards the other terminus of the molecule, is composed
mainly of a number of possibly homologous, repeating,
short peptides to each of which a chondroitin sulfate
chain is attached (Figures 1 and 3). Located between
clusters of these chondroitin sulfate containing peptides are short peptides to which keratan sulfate chains
are attached. This region which contains most of the
chondroitin sulfate and some of the keratan sulfate
chains is called the polysaccharide attachment region.
Between the hyaluronic acid binding region and the polysaccharide attachment region is a region containing
mainly keratan sulfate bound to peptide, called the keratan sulfate-rich region.
    A low molecular weight protein, called link protein
is also a component of proteoglycan aggregates. Link
protein appears to bind simultaneously to hyaluronate
and to the hyaluronic acid binding region of core protein, and stabilizes the bond between proteoglycan

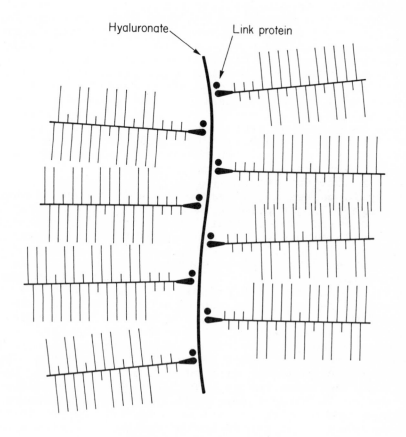

PROTEOGLYCAN AGGREGATE

Figure 3. Diagram of a cartilage proteoglycan aggregate

monomer and hyaluronate.

Isolation of Proteoglycan Species. Proof of the
concepts described above required that proteoglycan aggregate, proteoglycan monomer and link protein be isolated and characterized, and that the interactions of
proteoglycan monomer and link protein with hyaluronate
be studied by a variety of methods. The procedure now
generally used for the isolation of proteoglycan aggregates and proteoglycan monomer from cartilages involves
four steps called 1) dissociative extraction; 2) reassociation; 3) equilibrium density gradient centrifugation under associative conditions; and 4) equilibrium
density gradient centrifugation under dissociative conditions. In step 1, dissociative extraction, fresh wet
tissue is slowly stirred at 4° in 4 M guanidine hydrochloride (GnHCl), pH 5.8 to 6.3. The non-covalent
bonds between proteoglycan monomers, hyaluronate and
link protein are broken in concentrated solutions of
GnHCl or divalent cations (20-26). Proteoglycan monomer, hyaluronate, and link protein diffuse out of the
insoluble collagen network at a relatively rapid rate,
into the extraction solvent. The extract is separated
from the insoluble collagenous cartilage residue by
filtration. The filtered extract contains extraneous
proteins including proteases which must be separated
from the proteoglycans. In step 2, reassociation, proteoglycan monomer, link protein and hyaluronate are re-
associated into proteoglycan aggregates by dialyzing
off the GnHCl. In step 3, extraneous matrix proteins
and proteases are separated from the proteoglycan aggregates by an equilibrium density gradient centrifugation in 3.5 M CsCl, under associative conditions. The
gradient is frequently divided into six equal fractions.
The fractions from the top of this associative gradient
are called A1 through A6 (16,24,27). Proteoglycan aggregates are of high buoyant density and are concentrated in fraction A1 in the bottom one-sixth ($\rho \approx 1.6$ g/ml)
of the gradient. Proteoglycan fraction A1 is the preparation used for the physical characterization of proteoglycan aggregates by sedimentation velocity studies
(26-28), by electron microscopy (19), and as the starting point for the preparation of link protein (29-34)
or the hyaluronic acid-binding region of proteoglycan
monomer core protein (18,28). In step 4, equilibrium
density gradient centrifugation under dissociative conditions, the proteoglycan aggregate is separated into
its component species. Fraction A1 from an associative
gradient, which contains proteoglycan aggregate, is
dissolved in guanidine hydrochloride. The aggregate is dis-

sociated into proteoglycan monomer, hyaluronate and
link protein. Cesium chloride is added and a dissoci-
ative gradient is carried out in 4 $\underline{M}$ GnHCl-3 $\underline{M}$ CsCl.
The gradient is divided into six fractions called A1-D1
(bottom) through A1-D6 (top). Link protein is separated into A1-D6 at the top of the gradient. Hyaluronate
distributes in the middle of the gradient. Most of the
proteoglycan monomer, of high molecular weight and high
chondroitin sulfate-protein ratio, is concentrated at
the bottom of the gradient (A1-D1), free of hyaluronic
acid and link protein (35-37). However smaller amounts
of proteoglycan monomers of lower molecular weight dis-
tribute throughout the dissociative gradient, individual members of the polydisperse population of proteoglycan subunits banding at buoyant densities determined
mainly by their chondroitin sulfate to protein ratios.

Structural Basis for the Polydispersity of Proteoglycan Monomer. Polydisperse proteoglycan monomers from
bovine articular cartilage (36) and from bovine nasal
cartilage (37) have been separated into a series of
relatively monodisperse fractions by dissociative equilibrium density gradient centrifugation, and these
fractions have been chemically and physically characterized. Columns 2 through 9 of Table I show the chemical
composition and sedimentation coefficients of eight relatively monodisperse proteoglycan monomer fractions
from bovine articular cartilage (36). Column 1 of
Table I gives the amino acid composition of the hyaluronic acid-binding region of proteoglycan monomer isolated from bovine nasal cartilage by Heinegard and
Hascall (18). The molecular weight of the proteoglycan monomer increases in proportion to its chondroitin
sulfate content, as indicated by the increase in uronate
or galactosamine values with increasing size. The chondroitin sulfate-to-protein ratio also increases with
size. This relationship suggests that proteoglycan
monomers might contain protein cores identical in molecular weight and composition, to which chondroitin sulfate chains of different chain lengths are attached.
However, several observations rule out this possibility.
First, electron microscopic studies show that proteoglycan monomer core protein is of variable length (19).
Second, the amino acid composition of proteoglycan monomer varies in a characteristic fashion with molecular
weight. Proteoglycan monomer of the lowest molecular
weight (Column 2, Table I) contains little chondroitin
sulfate, and an amino acid composition relatively low
in serine and glycine, and relatively high in cysteine,
methionine and aspartic acid. As the molecular weight

TABLE I

Chemical composition and sedimentation coefficients (s_{20}^0) of proteoglycan monomer fractions from bovine articular cartilage (36). Column 1 shows the amino acid composition of the hyaluronic acid-binding region of proteoglycan monomer core protein, isolated by Heinegard and Hascall (<u>18</u>).

Column	1	2	3	4	5	6	7	8	9
FRACTIONS	HA-PGS*	A1-D3,4,5				A1-D2		A1-D1	
Yield, g/g		.019	.039	.036	.045	.074	.053	.209	.451
Uronate, %		9.7	10.3	11.5	15.3	16.1	17.1	19.0	20.1
Galactosamine		6.6	8.1	12.7	14.3	15.4	14.8	17.5	18.7
Hexose		12.5	13.5	12.9	14.3	13.3	11.7	11.8	12.2
Glucosamine		10.4	11.1	10.0	9.1	8.6	5.6	6.5	6.0
Sialate		3.0	3.1	2.9	2.4	2.8	1.8	1.8	1.4
Protein		30.7	23.9	17.3	13.0	14.9	10.3	11.1	9.9
Density, g/ml		1.41	1.46	1.52	1.61	1.57	1.57	1.65	1.62
s_{20}^0, subunit		5.7	7.8	8.8	9.7	10.3	10.8	12.7	14.3
s_{20}^0, aggregate			18.8	32.1					
			Amino Acid Composition residues/1000						
Aspartic acid	98	96	92	71	68	70	62	65	60
Threonine	60	61	65	·68	63	65	62	62	61
Serine	72	69	77	90	105	103	115	123	125
Glutamic acid	122	139	138	149	147	141	150	146	150
Proline	75	84	96	101	111	° 110	104	105	101
Glycine	80	81	87	93	102	102	117	114	118
Alanine	85	75	76	77	74	76	71	73	70
Half-cystine	21	20	21	17	14	17	12	13	12
Valine	60	60	56	56	59	56	59	59	57
Methionine	14	12	10	10	6	8	6	7	5
Isoleucine	48	35	34	33	32	31	32	33	40
Leucine	70	81	78	74	73	74	74	78	78
Tyrosine	48	42	27	33	29	20	27	25	24
Phenylalanine	33	40	45	41	43	41	39	38	38
Lysine	24	32	28	24	19	19	15	15	13
Histidine	14	14	17	12	11	11	12	13	13
Arginine	58	58	55	51	44	47	42	41	37

*HA-PGS: Hyaluronic acid binding region of PGS core protein, isolated by Heinegard and Hascall

and chondroitin sulfate content of the proteoglycan
monomer increases, there is a parallel increase in serine and glycine contents, and a decrease in cysteine,
methionine and aspartic acid contents of the proteoglycan monomer core protein. An interpretation of the
significance of these changes was made possible when
Heinegard and Hascall isolated and characterized the
hyaluronic acid-binding region of core protein (18).
     Heinegard and Hascall made a remarkable observa-
tion. They found that in proteoglycan aggregates, the
polysaccharide attachment region of core protein was
readily and selectively degraded by trypsin (18). As
shown in Figure 4, when proteoglycan aggregate was
treated with trypsin and chondroitinase, the polysaccharide attachment region was shattered into small
fragments. However, the central portion of the proteoglycan aggregate remained relatively unaltered. The
central portion of the proteoglycan aggregate consisted
of the hyaluronic acid-binding region of core protein,
non-covalently associated with link protein and hyaluronate. As shown in Figure 4, this complex, consisting of the hyaluronic acid-binding region, link protein
and hyaluronate, was separated from polysaccharide attachment region fragments by Sepharose 6B chromatography. The hyaluronic acid-binding region was then isolated from the complex by chromatography on Sephadex
G-200 in 4 $\underline{M}$ GnHCl. On sodium dodecyl sulfate-poly-
acrylamide gel electrophoresis, with or without mercap-
toethanol, the hyaluronic acid-binding region consisted
of a single polypeptide fragment approximately 90,000
in molecular weight.
     As shown in Table I, proteoglycan monomer of the
lowest molecular weight (Column 2) contains little
chondroitin sulfate, is relatively rich in keratan sulfate, and has an amino acid composition low in serine
and glycine, and high in cysteine, methionine and aspartic acid, almost identical to that of the hyaluronic
acid-binding region (Column 1, Table I). Proteoglycan
monomer of the lowest molecular weight appears to con-
sist mainly of the hyaluronic acid-binding region and
the keratan sulfate-rich region; it contains a short
polysaccharide attachment region composed of few Ser-
Gly containing peptides to which chondroitin sulfate
chains are attached (36,38,39). As the molecular
weight of proteoglycan monomer increases, the polysaccharide attachment region appears to progressively increase in length, with a concomitant increase in the
serine and glycine contents of core protein, and in the
chondroitin sulfate content of the monomer. This interpretation is in accord with electron microscopic

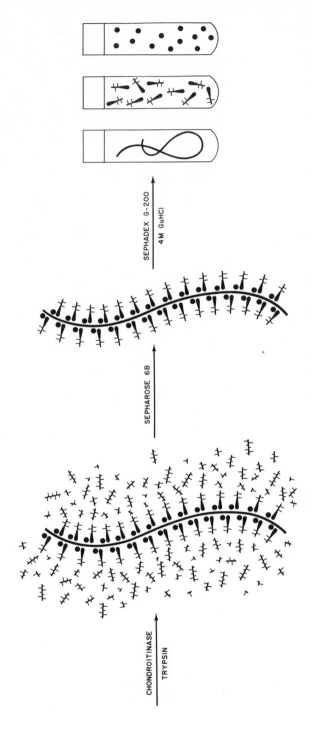

ISOLATION OF THE HYALURONIC ACID BINDING REGION OF PGS CORE PROTEIN

Figure 4. Diagramamtic representation of the procedure used by Heinegard and Hascall (18) for the isolation of the hyaluronic acid-binding region of proteoglycan monomer core protein

studies of proteoglycan aggregates, which show proteoglycan monomers varying in length from 1000 to 4000 Å
bound to hyaluronic acid at one terminus (19). In proteoglycan aggregates, all proteoglycan monomers contain
functional hyaluronic acid-binding regions, but possess
polysaccharide attachment regions of variable length.
Proteoglycan monomers appear to contain a hyaluronic
acid-binding region of constant size and composition
located at one terminus of the molecule and a polysaccharide attachment region of variable length extending
towards the other terminus of the molecule. The polydispersity of proteoglycan monomers appears to be determined by the variable length of the polysaccharide
attachment region of core protein.

Evidence that the Hyaluronic Acid-Binding Region
is Located at the NH$_2$-terminus of Proteoglycan Monomer
Core Protein. Proteoglycan monomer (A1-D1) core protein is approximately 200,000 in molecular weight and
consists of three regions which differ in structure
and function. Proteoglycan workers have therefore examined the possibility that core protein consists of
more than one polypeptide chain. However, once secreted extracellularly, proteoglycan monomer core protein
appears to be a covalently-linked structure, not dis-
sociable into smaller units. Further, proteoglycan
monomer core protein appears to contain a single NH$_2$-
terminal amino acid. Choi, et al. (40) have recently
determined the NH$_2$-terminal amino acids of proteoglycan
monomers (A1-D1-D1) from bovine nasal cartilage, bovine
articular cartilage, and bovine fetal epiphyseal cartilage. Proteoglycan monomers were dansylated, treated
with chondroitinase ABC, and hydrolyzed; the dansylated
amino acids were then separated by chromatography on
polyamide sheets (Figure 5). Valine was the only NH$_2$-
terminal amino acid demonstrated in each of the proteoglycan monomers examined. Valine was also the only
NH$_2$-terminal amino acid demonstrated when the hyaluronic acid-binding region of proteoglycan monomer core
protein (kindly provided by Dr. Vincent Hascall) was
examined. These observations suggest that the hyaluronic acid-binding region is located at the NH$_2$-termi-
nus of proteoglycan monomer core protein. Proteoglycan
monomers of smaller size from bovine nasal cartilage
(A1-D2 through A1-D4) were also examined. These con-
sistently showed valine as a single NH$_2$-terminal amino
acid. These results support the concept of a hyaluronic acid-binding region of constant size and composition
located at the NH$_2$-terminus, and a polysaccharide attachment region of variable length, extending towards

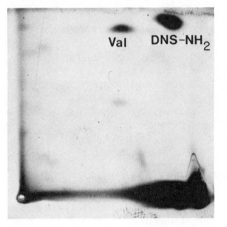

Figure 5. Photograph of a micropoly-
amide plate showing that valine is the
NH_2-terminal amino acid of proteoglycan
monomer. Identical results were obtained with the hyaluronic acid-binding
region of proteoglycan monomer core
protein.

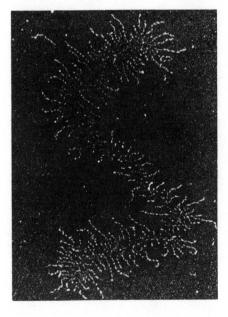

Journal of Biological Chemistry

Figure 6. Dark field electron micrograph of a large proteoglycan aggregate
from bovine articular cartilage. Proteoglycan monomers of varying length
arise laterally from the opposite sides
of an elongated central filament (hyaluronate) approximately 42,000 Å in
length ($\times 71,000$). From Rosenberg, et
al., J. Biol. Chem. (1975) 250, 1887–1883.

the COOH-terminus of proteoglycan monomer core protein.

The Proteoglycan Monomer-Hyaluronate Interaction.
Evidence that Hyaluronic Acid is the Filamentous Backbone of Proteoglycan Aggregates. Even in the absence
of link protein, proteoglycan monomers bind avidly to
hyaluronate to form high molecular weight complexes.
Electron micrographs of proteoglycan aggregates show
that proteoglycan monomers arise from an elongated central filament. Taken together, the results of chemical
binding studies and electron microscopic studies have
firmly established that hyaluronate forms a filamentous
backbone (Figure 3) to which monomers are non-covalently bound in proteoglycan aggregates.

Hardingham and Muir first demonstrated that the
addition of small amounts of hyaluronic acid to proteoglycan monomer resulted in the formation of high molecular weight complexes, demonstrable by gel chromatography or viscometry (11). These results were surprising, since it was thought that no hyaluronic acid was
present in cartilage. Hardingham and Muir quickly resolved the matter by isolating hyaluronate from pig
laryngeal cartilage, and from proteoglycan aggregates
prepared from this tissue (12). Hascall and Heinegard
(16) subsequently isolated hyaluronate from proteoglycan aggregates prepared from bovine nasal cartilage.
Hascall and Heinegard also showed that proteoglycan
monomer from bovine nasal cartilage, treated with
chondroitinase ABC to remove chondroitin sulfate, interacted with hyaluronate to form complexes of higher
molecular weight (17).

The molecular architecture of proteoglycan aggregates has been demonstrated by electron microscopy of
proteoglycan-cytochrome c monolayers (19). Figure 6
shows a dark field electron micrograph of a proteoglycan aggregate from bovine articular cartilage in which
proteoglycan monomers arise at fairly regular intervals
from the opposite sides of a central filament approximately 42,000 Å in length. Measurements of electron
micrographs of proteoglycan aggregates indicate that
the spacing between proteoglycan monomers along the
central filament is 200 to 300 Å (19). Calculations
made from the results of chemical binding studies indicate that the spacing between native proteoglycan
monomers on hyaluronate is $\sim$240 Å (13). This corres-
pondence between the spacing of monomers calculated
from chemical binding studies, and that demonstrated
by electron microscopy, supports the concept that the
filamentous backbone of the aggregate seen in electron
micrographs is hyaluronate (19).

Competitive Inhibition of the Proteoglycan Monomer-
Hyaluronate Interaction by Hyaluronate Oligosaccharides.
Proteoglycan monomers bind preferentially to hyaluronate
molecules of low molecular weight in the presence of
hyaluronate molecules of both high and low molecular
weights. Because of this, high molecular weight complexes formed by the binding of proteoglycan monomers
to high molecular weight hyaluronate are dissociated by
hyaluronate oligosaccharides (13,17). The competitive
inhibition of the proteoglycan monomer-hyaluronate interaction by hyaluronate oligosaccharides has been studied by Hascall and Heinegard (17) and by Hardingham
and Muir (13) with interesting results.

Hascall and Heinegard (17) prepared a proteoglycan
monomer core preparation of molecular weight 450,000 by
chondroitinase digestion of proteoglycan monomer. The
core preparation chromatographed as a single retarded
peak on Sepharose 2B. Mixtures of the core preparation
and hyaluronate of molecular weight 230,000 were prepared and chromatographed on Sepharose 2B. When a mixture of core and 20% hyaluronate were chromatographed,
there was little change in the elution volume of the
core. Under these conditions of excess hyaluronate,
only a few core molecules bound to each hyaluronate
molecule, and the complexes formed were relatively
small. Mixtures of core plus 4.1%, 1.7% and 0.8% hyaluronate were chromatographed; there was a progressive increase in the amount of core eluted in the void volume,
until at 0.6% hyaluronate approximately 70% of the core
molecules eluted near the void volume. Under these conditions, each hyaluronate molecule had been saturated
with core molecules, and eluted near the void volume as
a high molecular weight complex. Hascall and Heinegard
calculated that each core molecule, from which chondroitin sulfate chains have been removed with chondroitinase, occupied a length of about 8–10 hyaluronic acid disaccharides (80–100 Å) along the hyaluronic acid chain.
As noted above, in the intact aggregate (19), or in
complexes formed between hyaluronate and proteoglycan
monomers with intact chondroitin sulfate chains, the
minimum spacing between proteoglycan monomers is 250 Å.
These observations indicate that the spacing of proteoglycan monomers on hyaluronate is determined by the
lengths of chondroitin sulfate chains.

Oligosaccharides ranging from two to six repeating
units were prepared by testicular hyaluronidase digestion of hyaluronate, and isolated by Sephadex G-50 chromatography. Experiments were carried out to determine
the minimum chain length of hyaluronate to which core
molecules would bind. When core molecules were mixed

with high molecular weight hyaluronate, complexes were
formed which eluted in the void volume of Sepharose 2B
columns. When core molecules were mixed first with
hyaluronate octasaccharide, then with high molecular
weight hyaluronate, high molecular weight complexes
were still formed which eluted in the void volume. However, when core molecules were mixed first with hyaluronate decasaccharides, then with high molecular weight
hyaluronate, no high molecular weight complexes were
formed and the core molecules were retarded. Moreover,
the hyaluronate decasaccharides, which eluted in the
column total volume when chromatographed alone, now
eluted together with the core molecules. Similar results were obtained when the core molecules and high
molecular weight hyaluronate were mixed first, then
hyaluronate decasaccharides were added. Thus, proteoglycan monomer will bind to hyaluronate with a chain
length of five repeating units but not to hyaluronate
of four repeating units. Five repeating units may be
required for hyaluronate to assume a conformation essential for proteoglycan monomer-hyaluronate interaction. Core molecules did not bind to chondroitin oligosaccharides, which contain galactosamine, and differ
from hyaluronate only in that the 4-hydroxyl of the
amino sugar is in an axial rather than an equatorial
position.

Hardingham and Muir (13) have also studied the
binding of oligosaccharides of hyaluronate to proteoglycan monomer from pig laryngeal cartilage, using
viscometry. When oligosaccharides of appropriate chain
length were added to a proteoglycan monomer-hyaluronate
mixture, there was a decrease in viscosity that was
proportional to the amount of oligosaccharide added.
Relatively little effect was observed with hyaluronate
tetrasaccharide, hexasaccharide or octasaccharide, indicating that proteoglycan monomer does not bind to
these oligosaccharides. Decasaccharides and duodeca-
saccharides strongly inhibited the binding of proteoglycan to high molecular weight hyaluronic acid, resulting in a sharp decrease in viscosity.

Chemical Modifications of Hyaluronic Acid. Native
proteoglycan aggregates, and complexes formed between
proteoglycan monomer and hyaluronate, are dissociated
at pH 3 to 4, at which the carboxyl groups of hyaluronate are protonated. This observation suggested that
the carboxyl groups of hyaluronate might be involved
in the binding of proteoglycan monomer to hyaluronate.
Christner, Brown and Dziewiatkowski have recently studied the effect of chemical modification of the car-

boxyl groups of hyaluronate on the proteoglycan monomer-
hyaluonate interaction (41). Their results show that,
for proteoglycan monomer to bind to hyaluronate, hyaluronate carboxyl groups must be present and in a specific spatial orientation. Hyaluronate oligosaccharides
of 5 to 15 repeating units were prepared. These inhibited the binding of proteoglycan monomer to high molecular weight hyaluronate. The hyaluronate carboxyl
groups were then chemically modified in several ways.
First, the carboxyl groups were treated with diazometh-
ane to form the carboxymethyl esters. The oligosaccharides no longer inhibited the binding of proteoglycan
monomer to high molecular weight hyaluronate. When the
methyl groups were removed by saponification, the hyaluronate oligosaccharides again inhibited the binding
of proteoglycan monomer to high molecular weight hyal-
ronate. If the carboxymethyl ester was reduced with
NaBH₄, so that the glucuronic acid residue of hyaluronate was transformed to glucose, the oligosaccharide
also lost its inhibitory capacity.

     In other experiments, the amide was formed between
glycine methyl ester and the carboxyl groups of glucuronic acid residues. The inhibitory capacity of the
oligosaccharides was lost. The methyl group was removed by saponification, yielding the glycine amide of
glucuronic acid, in which a free carboxylate group
arises from glucuronic acid, but is displaced by the
interposition of a glycine residue. The inhibitory
capacity of the oligosaccharide was not restored. The
conformation of the glucuronic acid carboxylate group
is essential for proteoglycan monomer to bind to hyaluronate.

     Structure and Function of Link Protein. Cartilage
proteoglycan aggregates are formed by the non-covalent
association of proteoglycan monomer, hyaluronate and
link protein. Link protein can be isolated from proteoglycan aggregate by equilibrium density gradient centrifugation under dissociative conditions, followed by
gel chromatography under dissociative conditions. Link
protein is first partially purified by equilibrium density gradient centrifugation of proteoglycan fraction
A1 in 4 M GnHCl - 3 M CsCl. Link protein is recovered
at low buoyant densities from the top one-third of 4 M
GnHCl - 3 M CsCl gradients, mixed with some hyaluronate
and low molecular weight, protein-rich proteoglycan
monomer. Most of the hyaluronate can be separated from
link protein by a sequential 4 M GnHCl - 2 M CsCl dissociative gradient. Link protein may then be separated
from low molecular weight, protein-rich proteoglycan

monomer by chromatography on Sephadex G-200 in 4 $\underline{M}$
GnHCl (18), or on Sephacryl S-200 in 4 $\underline{M}$ GnHCl, or on
Ultragel 34 in 1% SDS (30) or on Sepharose CL-6B in 4
$\underline{M}$ GnHCl (33). In our hands, chromatography on Sephacryl S-200 in 4 $\underline{M}$ GnHCl yields functionally active,
pure link protein, free of protein-rich proteoglycan
monomer, based on immunodiffusion studies (34).

Link protein preparations from most cartilages,
isolated as described above, contain two proteins
approximately 44,000 and 48,000 in molecular weight, on
sodium dodecyl sulfate polyacrylamide gel electrophoresis. These two proteins have been called link proteins 1 and 2. Molecular weights of link proteins 1
and 2 from several cartilages are given in Table II.

TABLE II

Molecular Weights (x10^3) of Link Proteins from Several
Cartilages, Based on Sodium Dodecyl Sulfate Polyacrylamide Gel Electrophoresis.

CARTILAGE	LINK PROTEIN 1	2	REFERENCE
Bovine Nasal	45	40	Oegema, et al. (28)
Bovine Nasal	48	44	Tang, et al. (34)
Bovine Nasal	51	47	Baker and Caterson (30,33)
Human Articular	49	40	Pal, et al. (27)
Human Chondrosarcoma	49	40	Pal, et al. (27)
Swarm Rat Chondro-sarcoma	--	40	Oegema, et al. (29)

Baker and Caterson have separated link proteins 1 and 2
by preparative gel electrophoresis, and have presented
evidence that link proteins 1 and 2 are glycoproteins
which differ in their oligosaccharide components, but
not in their amino acid composition (33). The compositions of link proteins 1 and 2 from bovine nasal cartilage (kindly provided by Dr. John Baker) are shown
in Table III (33).

Close examination of the model of the proteoglycan
aggregate depicted in Figure 3 suggests that link protein binds simultaneously to hyaluronate, and to the
hyaluronic acid-binding region of proteoglycan monomer
core protein, and may serve to strengthen or stabilize
the binding of proteoglycan monomer to hyaluronate.
The studies of Heinegard and Hascall provide evidence
that link protein is located in the region where the
hyaluronic acid-binding region of proteoglycan monomer

TABLE III

Chemical Compositions of Link Proteins 1 and 2

Link Protein 1   Link Protein 2
AMINO ACID COMPOSITION
(residues per 1000 residues)

	Link Protein 1	Link Protein 2
L-aspartic acid	135	133
L-threonine	52	52
L-serine	62	63
L-glutamic acid	76	84
L-proline	48	54
Glycine	104	103
L-alanine	80	77
L-valine	61	62
L-methionine	3	2
L-isoleucine	29	28
L-leucine	80	82
L-tyrosine	66	61
L-phenylalanine	53	52
L-histidine	29	26
L-lysine	58	59
L-arginine	64	62

CARBOHYDRATE COMPOSITION
(moles per 10^5 g protein)

	Link Protein 1	Link Protein 2
L-fucose	1.8	2.3
D-mannose	16.6	5.7
D-galactose	5.7	1.1
N-acetyl-D-glucosamine	16.5	5.1
N-acetyl-D-galactosamine	5.6	1.3
Sialic acid	0.9	tr
Total	47.2	15.6
Total (% by weight)	9.5	3.0

(FROM: Baker, J.R. and Caterson, B.: The Isolation and Charac-
terization of the Link Proteins from Proteoglycan Aggregates of
Bovine Nasal Cartilage. Submitted for publication in J. Biol.
Chem.)

binds to hyaluronate; as noted above (Figure 4), degradation of proteoglycan aggregates with trypsin removes the polysaccharide attachment region of proteoglycan monomers, yielding a complex composed of hyaluronate, hyaluronic acid-binding region of proteoglycan
monomer core protein, and link protein, representing
the central portion of proteoglycan aggregate.

Caterson and Baker (31) have shown that link protein binds to proteoglycan monomer in the absence of
hyaluronate. Link protein from bovine nasal cartilage
was isolated by chromatography on Sepharose CL 6B in
4 M GnHCl. The link protein was mixed with proteoglycan monomer. The mixture was subjected to equilibrium
density gradient centrifugation under associative conditions. The distribution of link protein in the gradient followed that of proteoglycan monomer, which distributed at high buoyant densities. When the link pro-
tein-proteoglycan monomer mixture was chromatographed
on Sepharose CL 2B under associative conditions, link
protein was eluted with proteoglycan monomer. These
observations indicate that link protein non-covalently
associates with proteoglycan monomer in the absence of
hyaluronate.

Tang, et al. (34) have provided evidence that link
protein stabilizes the binding of proteoglycan monomer
to hyaluronate. Link protein present in low density
fractions from the top of 4 M GnHCl-2 M CsCl gradients
was separated from protein-rich proteoglycan monomer
and hyaluronate by chromatography on Sephacryl S-200
in 4 M GnHCl. Link protein prepared by this procedure
was immunologically pure. Since link protein is insoluble in most associative solvents, a study was carried
out to identify associative solvents in which link protein is soluble. Several associative solvents were
identified in which link protein is soluble. In these
solvents, link protein was present as an 8 S species in
sedimentation velocity studies, suggesting that link
protein exists as an oligomer under associative conditions. Proteoglycan monomer was prepared from bovine
nasal cartilage, which interacted with hyaluronate in
the absence of link protein, to form a high molecular
weight complex (s_{20}^0 = 68 S) demonstrable in sedimentation velocity studies. The addition of link protein to
the monomer-hyaluronate mixture resulted in an increase
in the sedimentation coefficient of the complex from
68 to 81 S.

The complex formed between proteoglycan monomer
and hyaluronate in the absence of link protein was un-
stable at acid pH. Approximately one-half of the complex was dissociated at pH 5; all of the complex was

dissociated at pH 4 or 3. The addition of link protein
greatly increased the stability of the complex against
dissociation at acid pH. In the presence of link protein, there was no detectable dissociation of the complex at pH 5, and the complex was only 50% dissociated
at pH 4.0. Even at a pH of 3, some complex remained un-
dissociated in the presence of link protein. These
observations indicate that one biologic role of link
protein is to stabilize the interaction between proteoglycan monomers and hyaluronate in proteoglycan aggregates.

Blood Vessel Proteoglycans

Cartilage contains one class of proteoglycan monomer in which chondroitin sulfate and keratan sulfate
are covalently bound to the same protein core. Blood
vessel, kidney and lung contain several glycosaminoglycans, including chondroitin 4-sulfate, chondroitin 6-
sulfate, dermatan sulfate, heparan sulfate and heparin
(42-53). Compared to cartilage, little is known about
the proteoglycans of blood vessel, kidney and lung.
However, blood vessel and lung parenchyma contain at
least two classes of proteoglycan monomers, different
from that found in cartilage. One class consists of
dermatan sulfate and chondroitin sulfate chains bound
to the same protein core. The other class consists of
heparan sulfate bound to a protein core. The study
of proteoglycans has recently entered an exciting new
phase, in which interest is being focused on the isolation and characterization of each of these classes
of proteoglycan monomers, their localization in the
intercellular matrix, basement membranes and plasma
membranes of cells in different tissues, and the elucidation of their biologic functions in blood vessel,
kidney and lung.

Dermatan Sulfate-Containing Proteoglycans. The
disaccharide repeating unit of dermatan sulfate (Figure
2) consists of L-iduronic acid and N-acetylgalactosamine (58). L-iduronic acid is the C5 epimer of D-gluc-
uronic acid, in which the carboxyl group is in an axial
rather than in an equatorial position. The N-acetylgalactosamine residue of dermatan sulfate carries an ester sulfate group usually on carbon number 4, but sometimes on carbon number 6.
Dermatan sulfate chains contain repeating units
composed of glucuronic acid and N-acetylgalactosamine
(Figure 2) as well as iduronic acid and N-acetylgalactosamine. Thus, the dermatan sulfate chain is a co-

polymer composed of dermatan sulfate and chondroitin
sulfate repeating units. The hybrid structure of dermatan sulfate has been extensively studied by Fransson
(54-58). Dermatan sulfate was isolated from different
sources by methods capable of removing chondroitin sulfate. The dermatan sulfate fractions contained glucuronic acid as well as iduronic acid. The dermatan
sulfate fractions were degraded with testicular hyaluronidase. This enzyme cleaves GalNAc($\beta1\rightarrow4$)GlcUA linkages in chondroitin sulfate repeating units, but not
GalNAc($\beta1\rightarrow4$)IdUA linkages in dermatan sulfate repeating
units. Viscosity measurements showed that dermatan
sulfate chains were degraded by testicular hyaluronidase. Fragments were formed with glucuronic acid residues at newly formed non-reducing termini. Following
testicular hyaluronidase degradation of umbilical cord
dermatan sulfate, Fransson isolated and characterized
a hybrid octasaccharide with the following structure:

$$GlcUA\rightarrow GalNAc\rightarrow IdUA\rightarrow GalNAc\rightarrow IdUA\rightarrow GalNAc\rightarrow GlcUA\rightarrow GalNAc$$
$$OSO_3^-\qquad OSO_3^-\qquad OSO_3^-\qquad OSO_3^-$$

When dermatan sulfate chains were degraded with testicular hyaluronidase, the fragments formed were mainly
of high molecular weight and composed of dermatan sulfate repeating units, or were low molecular weight
oligosaccharides, containing most of the glucuronic
acid. Fragments of intermediate molecular weight were
scarce. Fransson suggested that in the native dermatan
sulfate chains, long segments composed entirely of iduronic acid-containing repeating units are separated
by clusters of glucuronic acid-containing repeating
units, located adjacent to one another. Fransson's
observations were based on a study of dermatan sulfate
isolated from pig skin and human umbilical core. Little is known about the co-polymeric structure of dermatan sulfate from blood vessel, kidney and lung.

Blood vessel, kidney and lung contain proteoglycan
monomers consisting of dermatan sulfate and chondroitin
sulfate chains covalently bound to the same protein
core. The linkage region of dermatan sulfate to protein is identical to that of chondroitin sulfate (Figure 2) (58, 59). Only a few attempts have been made
to isolate and characterize dermatan sulfate containing
proteoglycan monomers. In 1971, Kresse, Heidel and
Buddecke (60, 61) extracted proteoglycans from bovine
aorta by high speed homogenization in 0.15 $\underline{M}$ phosphate,
0.05 $\underline{M}$ EDTA, pH 7. Dermatan sulfate-containing proteoglycan was purified by repeated cetylpyridinium chloride precipitates with $MgCl_2$. A dermatan sulfate-con-

taining proteoglycan was obtained which contained 20%
protein and 80% glycosaminoglycan. The glycosaminoglycan consisted of 75% chondroitin sulfate and 25%
dermatan sulfate. The authors stated that the proteoglycan behaved as a single component in the analytical
ultracentrifuge, although no data from sedimentation
velocity experiments or the conditions of the sedimentation velocity studies were presented. The weight
average molecular weight of the dermatan sulfate containing proteoglycan from light scattering was 2×10^6.
Following degradation of the dermatan sulfate-containing proteoglycan with pronase and testicular hyaluronidase, oligosaccharides ranging from tetrasaccharides to
octasaccharides were isolated, whose iduronic acid/gluc-
uronic acid ratio increased with increasing chain length.
     Studies of blood vessels and other tissues, and
of cells in culture indicate that dermatan sulfate-
containing proteoglycans are distributed throughout the
intercellular substance interconnecting collagen fibers
elastin and cells, while heparan sulfate-containing
proteoglycans are found in the plasma membranes of
cells. Wight and Ross recently studied the ultrastructural localization of proteoglycans in the intima of
non-human primate arteries (62). Numerous 200-500 Å
diameter polygonal granules with a marked affinity for
ruthenium red were distributed throughout the intercellular substance (Figure 7). The granules possessed
30-60 Å thick filamentous projections which appeared to
interconnect adjacent granules. The granules and their
filaments interconnected collagen fibers at regular intervals in register with the periodicity of the collagen fibers, and elastic fibers, and appeared to form
connections between the plasma membranes of smooth muscle cells and intercellular fibers (Figure 7). Most
of the intercellular granules and filaments were removed with chondroitinase ABC. Wight and Ross (63)
also found that 60-80% of the glycosaminoglycan synthesized and secreted into the medium by arterial smooth
muscle cells in culture was dermatan sulfate, while
only 10-20% was chondroitin 4- or 6-sulfate. Taken together, the results indicated that most of the intercellular matrix granules are probably dermatan sulfate-
containing proteoglycans. Wight and Ross suggested
that the intercellular proteoglycan might function to
hold collagen fibers, elastin and cells together, and
at the same time maintain tissue turgor as a result of
their elastic properties. They suggested that the proteoglycans might function as a type of plastic inter-
stitial substance, important in absorbing and/or dissi-
pating stress.

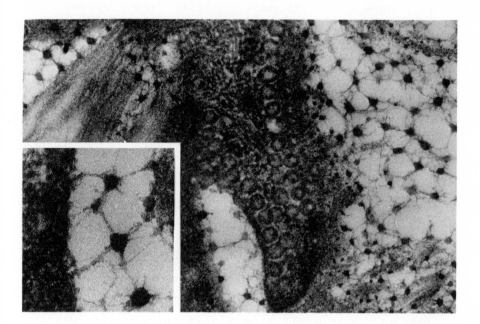

Journal of Cell Biology

Figure 7. Portion of an intimal smooth muscle cell and the adjacent intercellular matrix from a primate artery (Macaca nemestrina) stained with ruthenium red. The ruthenium red stains numerous polygonal granules associated with each other through 30–60 Å diameter filamentous projections which, based on studies with chondroitinases, must represent dermatan sulfate–containing proteoglycans (×40,000). As shown in the insert, the 30–60 Å filaments interconnect granules, collagen fibers, elastin, and the surfaces of cells (×140,000). Reproduced from Proteoglycans in Primate Arteries. I. Ultrastructural Localization and Distribution in the Intima, by Thomas N. Wight and Russell Ross, J. Cell Biol. (1975) 67, 660–674.

Eisenstein and Kuettner reported similar observations in an electron microscopic study of the ultra-
structure of proteoglycans in bovine aorta (64,65). Ruthenium red stained granules, 300-400 Å in diameter,
interconnected by slender filaments, were found in the
intercellular matrix, on the surfaces of collagen fibers, at the edges of elastic fibers and near the plasma membranes of smooth muscle cells. The granules dis-
appeared after digestion with chondroitinase ABC, or
after extraction with 4 M guanidine hydrochloride. The
presence of dermatan sulfate-containing proteoglycans
on the surfaces of collagen fibers was also indicated
by the binding of peroxidase-labeled antibodies raised
against dermatan sulfate-containing proteoglycans to
the surfaces of collagen fibers.
    In both the studies of Wight and Ross, and of
Eisenstein and Kuettner, ruthenium red staining material was also found in the plasma membranes of endothelial cells and smooth muscle cells which was not removed with chondroitinase ABC and is therefore not chondroitin sulfate or dermatan sulfate.
    Observations by several workers suggest that at
least some of the ruthenium red staining material found
in the plasma membranes of cells is heparan sulfate-
containing proteoglycan. Cells in culture appear to
synthesize three discreet pools of glycosaminoglycans;
1) an extracellular pool, secreted into the culture
medium, probably representative of the proteoglycans
secreted into the intercellular matrix in vivo; 2) a
cell-membrane associated or pericellular pool; and 3)
an intracellular pool. For a particular cell type,
each pool frequently shows a characteristic distribution in terms of the percentage of the total glycosaminoglycan that is chondroitin sulfate, dermatan sulfate,
heparan sulfate, heparin or hyaluronate. Kraemer has
shown that heparan sulfate is present as a proteoglycan
in the plasma membrane of a variety of cells (66-70).
Treatment of cultured cells with trypsin released fragments of plasma membrane heparan sulfate-containing
proteoglycan with a molecular weight of 135,000 (70).
The fragments were excluded on Bio-Gel A 0.5 m. When
the exculded material was treated with nitrous acid,
oligosaccharides characteristic of heparan sulfate were
formed, which were retarded on Bio-Gel P-10. After
alkaline borohydride treatment of the tryptic fragments,
reisolated heparan sulfate chains had a molecular weight
of 44,000, suggesting at least 3 chains per tryptic
fragment. Silbert has studied similar heparan sulfate-
containing tryptic fragments derived from the plasma
membranes of cultured normal skin fibroblasts (71,72).

Using heparinase and heparitinase from Flavobacterium
heparinum, Buonassisi has provided evidence for the
presence of heparan sulfate-containing proteoglycans
in the plasma membranes of endothelial cells from rabbit aorta (73,74).

The observations described above raise questions
about the functions of dermatan sulfate-containing
proteoglycans in the intercellular matrix of arterial
wall, and about the functions of heparan sulfate-containing proteoglycans in the plasma membranes of endothelial cells and smooth muscle cells. Essentially
nothing is known about the biological function of heparan sulfate-containing proteoglycans in the plasma
membranes of cells. Methods for the extraction and
isolation of native heparan sulfate-containing proteoglycans have yet to be developed.

However, some functions of dermatan sulfate-containing proteoglycans are gradually becoming apparent.
As indicated above, the ultrastructural studies of
Wight and Ross and of Eisenstein and Kuettner suggest
that dermatan sulfate-containing proteoglycans interconnect collagen, elastin and cells, and contribute to
the elastic and mechanical properties of arterial wall.
Implicit in this concept is the idea that dermatan
sulfate-containing proteoglycans may bind to and non-
covalently associate with collagen. Recent studies
show that dermatan sulfate-containing proteoglycans
non-covalently bind to collagen, and influence both
the conformational stability of collagen monomer, and
the formation of collagen fibrils from collagen monomer. Toole and Lowther (75) isolated a dermatan sulfate containing proteoglycan, following extraction of
bovine heart valves with 6 M urea at 60°. When it was
mixed with collagen monomer, collagen fibrils with typical periodicity were formed immediately. No fibrils
were formed when the collagen monomer was mixed with
hyaluronate or chondroitin sulfate-containing proteoglycan. Toole and Lowther suggested that the primary
biologic role of dermatan sulfate-containing proteoglycans might be in the formation and orientation of
collagen fibrils from collagen monomer.

Blackwell and his co-workers have used circular
dichroism spectroscopy to study the interactions of
glycosaminoglycans with collagen, and with synthetic
cationic polypeptides. In the absence of glycosaminoglycans, poly-L-lysine and poly-L-arginine exist in
an extended charged coil conformation. Glycosaminoglycans bind to these cationic polypeptides and cause them
to assume an α-helical conformation. In a series of
systematic studies (76-83), Blackwell and his co-workers

have extensively examined the conformation-directing
effect of glycosaminoglycans on cationic polypeptides.
The extent of α-helix formation was assessed from the
negative ellipticity found on circular dichroism spectroscopy in dilute aqueous solution at neutral pH.
For each mixture, maximum α-helix content was obtained
at a characteristic ratio of amino acid residues to
disaccharide repeating units (Table IV).

TABLE IV
Interactions of Glycosaminoglycans with Cationic Polypeptides.*

Comparison of Conformation-Directing Effects, Residue
Ratios, and Melting Temperature at Maximum Interaction
for the Seven Glycosaminoglycans (76,81,83).

	HA	C4S	HS	C6S	KS1	DS	HEP
			Poly-L-arginine				
Ratio	1:1	2:1	1:1	2:1	1.2:1	1.4:1	3.3: 1
Effect	α	α	α	α	α	α	α
t_m (°C)	35.0	54.5	65.0	76.0	>90	>90	>90

	HA	C4S	HS	C6S	KS1	DS	HEP
			Poly-L-lysine				
Ratio	1:1	1:1	2:1	1:1	1.2: 1	1.4:1	2.3: 1
Effect	R	α	R	α	R	α	α
t_m (°C)	-	25.0	-	47.0	-	74.5	>90

* Abbreviations used in TABLE IV

HA - hyaluronic acid   C4S - chondroitin 4-sulfate
HS - heparan sulfate   C6S - chondroitin 6-sulfate
KS1 - keratan sulfate   DS - dermatan sulfate
                    HEP - heparin

Glycosaminoglycans were compared in terms of the residue ratio at which maximum α-helix formation was obtained (TABLE IV). Glycosaminoglycan-polypeptide interactions were weakened and finally abolished, as the
temperature was increased. The α-helical conformation
of the polypeptide was lost, and the polypeptide re-
verted to an extended charged coil conformation. The
melting temperature for each glycosaminoglycan-polypeptide mixture was defined as the midpoint of the
transition from α-helix to extended coil of the polypeptide. The melting temperature of a particular polypeptide was different for each glycosaminoglycan (Table IV). Therefore, the strength of the interaction
between glyc aminoglycans and cationic polypeptides
could also be evaluated in terms of the melting temperature (t_m, °C) of the mixture. As indicated by the
dermatan sulfate-poly-L-arginine or poly-L-lysine melt-

ing temperature shown in Table IV, the interaction between dermatan sulfate and cationic polypeptides is particularly strong. In connective tissues such as blood vessel, kidney and lung, dermatan sulfate-containing proteoglycans must interact strongly with collagen fibers, and increase the stability of the collagen monomer triple helix (84). In doing so, dermatan sulfate-containing proteoglycans must also envelop and shield collagen fibers, thereby regulating the interactions of collagen with platelets and other cells.

LITERATURE CITED

1. Gregory, J.D., Laurent, T., and Roden, L.,J. Biol. Chem. (1964) 239, 3312-3320.
2. Gregory, J.D., and Roden, L., Biochem. Biophys. Res. Comm. (1961) 5, 430-434.
3. Lindahl, U., and Roden, L., J. Biol. Chem. (1966) 241, 2113-2119.
4. Roden, L., and Armand, G., J. Biol. Chem. (1966) 241, 65-70.
5. Roden, L., and Smith, R., J. Biol. Chem. (1966) 241, 5949-5954.
6. Seno, N., Meyer, K., Anderson, B., and Hoffman, P. J. Biol. Chem. (1965) 240, 1005-1010.
7. Anderson, B., Hoffman, P., and Meyer, K., J. Biol. Chem. (1965) 240, 156-167.
8. Bray, B.A., Lieberman, R., and Meyer, K., J. Biol. Chem. (1967) 242. 3373-3380.
9. Hopwood, J.J., and Robinson, H.C., Biochem. J. (1974) 141, 57-69.
10. Kieras, F.J., J. Biol. Chem. (1974) 249, 7506-7513
11. Hardingham, T.E., and Muir, H., Biochim. Biophys. Acta (1972) 279, 401-405.
12. Hardingham, T.E., and Muir, H., Biochem. Soc.Trans. (Dublin) (1973) 1, 282-284.
13. Hardingham, T.E., and Muir, H., Biochem. J. (1973) 135, 905-908.
14. Hardinghan, T.E., and Muir, H., Biochem. J. (1974) 139, 565-581.
15. Hardingham, T.E., Ewins, R.J.F., and Muir, H., Biochem. J. (1976) 157, 127-143.
16. Hascall, V.C., and Heinegard, D., J. Biol. Chem. (1974) 249, 4232-4241.
17. Hascall, V.C., and Heinegard, D., J. Biol. Chem. (1974) 249, 4242-4249.
18. Heinegard, D., and Hascall, V.C., J. Biol. Chem. (1974) 249, 4250-4256.
19. Rosenberg, L., Hellman, W. and Kleinschmidt, A.

J. Biol. Chem. (1975) 250, 1877-1883.
20. Sajdera, S.W., and Hascall, V.C., J. Biol. Chem. (1969) 244, 77-87.
21. Hascall, V.C., and Sajdera, S.W., J. Biol. Chem. (1969) 244, 2384-2396.
22. Rosenberg, L., Pal, S., Beale, R., and Schubert, M., J. Biol. Chem. (1970) 245, 4112-4122.
23. Rosenberg, L., Hellmann, W., and Kleinschmidt, A., J. Biol. Chem. (1970) 245, 4123-4130.
24. Heinegard, D., Biochim. Biophys. Acta (1972) 285, 181-192.
25. Heinegard, D., Biochim. Biophys. Acta (1972) 285, 193-207.
26. Rosenberg, L., Pal, S., and Beale, R., J. Biol. Chem. (1973) 248, 3681-3690.
27. Pal, S., Strider, W., Margolis, R., Gallo, G., Lee-Huang, S., and Rosenberg, L., J. Biol. Chem. (1978) 253, 1279-1289.
28. Oegema, T.R., Jr., Hascall, V.C., and Dziewiatkowski, D.D., J. Biol. Chem. (1975) 250, 6151-6159.
29. Oegema, T.R., Jr., Brown, M., and Dziewiatkowski, D.D., J. Biol. Chem. (1977) 252, 6470-6477.
30. Baker, J., and Caterson, B., Biochem. Biophys. Res. Comm. (1977) 77, 1-10.
31. Caterson, B., and Baker, J., Biochem. Biophys. Res. Comm. (1978) 80, 496-503.
32. Caterson, B., and Baker, J., Submitted for publication (1978).
33. Baker, J., and Caterson, B., Submitted for publication (1978).
34. Tang, L.H., Rosenberg, L., and Poole, A.R., In preparation.
35. Hascall, V.C., and Sajdera, S.W., J. Biol. Chem. (1970) 245, 4920-4930.
36. Rosenberg, L., Wolfenstein-Todel, C., Margolis, R. Pal, S., and Strider, W., J. Biol. Chem. (1976) 251, 6439-6444.
37. Heinegard, D., J. Biol. Chem. (1977) 252, 1980-1989.
38. Baker, J.R., Roden, L., and Yamagata, S., Biochem. J. (1971) 125, 93P.
39. Baker, J.R., Roden, L., and Stoolmiller, A.C., J. Biol. Chem. (1972) 247, 3838-3847.
40. Choi, H.U., Tang, L.H., and Rosenberg, L., In preparation.
41. Christner, J.E., Brown, M.L., and Dziewiatkowski, D.D., Biochem. J. (1977) 167, 711-716.
42. Berenson, G.S., Biochim. Biophys. Acta (1958) 28, 176-183.
43. Zugibe, F.T., J. Histochem. Cytochem. (1962) 10,

441-447.
44. Zugibe, F.T., J. Histochem. Cytochem. (1962) 10. 448-461.
45. Kumar, V., Berenson, G.S., Ruiz, H., Dalferes, E.R., Jr., and Strong, J.P., J. Atheroscler. Res. (1967) 7, 573-581.
46. Kumar, V., Berenson, G.S., Ruiz, H., Dalferes, E.R., Jr., and Strong, J.P., J. Atheroscler. Res. (1967) 7, 583-590.
47. Dalferes, E.R., Jr., Ruiz, H., Radhakrishnamurthy B., and Berenson, G.S., Athersoclerosis (1971) 13, 121-131.
48. Berenson, G.S., Radhakrishnamurthy, B., Dalferes, E.R., Jr., and Srinivasan, S.R., Human Pathol. 2, 57-78.
49. Engel, U.R., Atherosclerosis (1971) 13, 45-60.
50. Murata, K., Nakazawa, K., and Hamai, A., Atherosclerosis (1975) 21, 93-103.
51. Nakazawa, K., and Murata, K., Paroi Arterielle (1975) 2, 302-211.
52. Sjoberg, I., and Fransson, L.A., Biochem. J. (1977) 167, 383-392.
53. Radhakrishnamurthy, B., Ruiz, H.A., and Berenson, G.S., J. Biol. Chem. (1977) 252, 4831-4841.
54. Fransson, L.A., and Roden, L., J. Biol. Chem. (1967) 242, 4161-4169.
55. Fransson, L.A., and Roden, L., J. Biol. Chem. (1967) 242, 4170-4175.
56. Fransson, L.A., J. Biol. Chem. (1968) 243, 1504-1510.
57. Fransson, L.A., Biochim. Biophys. Acta (1968) 156, 311-316.
58. Fransson, L.A., "Chemistry and Molecular Biology of the Intercellular Matrix" E.A. Balazs, Ed. p. 823, Academic Press, New York, (1970).
59. Bella, A., Jr., and Danishefsky, I., J. Biol. Chem. (1968) 243, 2660-2664.
60. Kresse, H., Heidel,H., and Buddecke, E., Eur. J. Biochem. (1971) 22, 557-562.
61. Buddecke, E., Kresse, H., Segeth, G. and Figura, K.V. "Connective Tissues. Biochemistry and Pathology" R. Fricke, and Hartmann, F., eds. Springer-Verlag, New York, (1974).
62. Wight, T., and Ross, R., J. Cell Biol. (1975) 67, 660-674.
63. Wight, T., and Ross, R., J. Cell Biol. (1975) 67, 675-686.
64. Eisenstein, R., Larsson, S.E., Kuettner, K.E., Sorgente, N., and Hascall, V.C., Atherosclerosis (1975) 22, 1-17.

65. Eisenstein, R., and Kuettner, K., Atherosclerosis
     (1976) 27, 37-46.
66. Kraemer, P.M., J. Cell Physiol. (1968) 71, 109-
     120.
67. Kraemer, P.M., Biochemistry (1971) 10, 1437-1445.
68. Kraemer, P.M., Biochemistry (1971) 10, 1445-1451.
69. Kraemer, P.M., J. Cell Biol. (1972) 55, 713-717.
70. Kraemer, P.M., and Smith, D.A., Biochem. Biophys.
     Res. Comm. (1974) 56, 423-430.
71. Kleinman, H.K., Silbert, J.E., and Silbert, C.K.,
     Conn. Tissue Res. (1975) 4, 17-23.
72. Silbert, J.E., Kleinman, H.K., and Silbert, C.K.,
     "Heparin. Structure, Function and Clinical Im-
     plication. Advances in Experimental Medicine and
     Biology" Vol. 52, Bradshaw, R.A., and Wessler, S.
     eds., pp. 51-60. Plenum Publishing Corporation
     New York, (1975).
73. Buonassisi, V., Exp. Cell Res. (1973) 76, 363-368.
74. Buonassisi, V., and Root, M., Biochim. Biophys.
     Acta (1975) 385, 1-10.
75. Toole, B.P., and Lowther, D.A., Arch. Biochem.
     Biophys. (1968) 128, 567-578.
76. Gelman, R.A. and Blackwell, J., Arch. Biochem.
     Biophys. (1973) 159, 427-433.
77. Gelman, R.A., Rippon, W.B., and Blackwell, J.
     Biopolymers (1973) 12, 541-558.
78. Gelman, R.A., Glaser, D.N., and Blackwell, J.
     Biopolymers, (1973) 12, 1223-1232.
79. Gelman, R.A., and Blackwell, J., Biopolymers,
     (1973) 12, 1959-1974.
80. Gelman, R.A., and Blackwell, J., Biochim. Biophys.
     Acta (1974) 342, 254-261.
81. Schodt, K.P., and Blackwell, J., Biopolymers,
     (1976) 15, 469-482.
82. Schodt, K.P., Gelman, R.A., and Blackwell, J.
     Biopolymers (1976) 15, 1965-1977.
83. Gelman, R.A., and Blackwell, J. Biopolymers,
     (1974) 13, 139-156.
84. Gelman, R.A. and Blackwell, J., Conn. Tissue
     Res. (1973) 2, 31-35.

Supported by grants AM HD 21498 and CA AM 23945 from
the National Institutes of Health

RECEIVED September 8, 1978.

INDEX

INDEX